THE RETURN OF THE GODS

Erich von Däniken came to prominence in the 1970s with his bestseller *Chariots of the Gods?* and has since written many books on a similar theme. He continues his research from his home in Switzerland and remains a passionate advocate of the reality of extraterrestrial visits.

by the same author

Chariots of the Gods?

The Return of the Gods

EVIDENCE OF EXTRATERRESTRIAL VISITATIONS

Erich von Däniken

ELEMENT
Shaftesbury, Dorset • Rockport, Massachusetts
Brisbane, Queensland

First published in Great Britain in 1997 by
Element Books Limited, Shaftesbury, Dorset SP7 8BP

Published in the USA in 1997 by
Element Books, Inc.
PO Box 830, Rockport, MA 01966

Published in Australia in 1997 by
Element Books Limited
for Jacaranda Wiley Limited
33 Park Road, Milton, Brisbane 4064

Translated by Matthew Barton
Cover illustration courtesy of Telegraph Colour Library
Cover design by Mark Slader
Design by Linda Reed and Joss Nizan
Typeset by Bournemouth Colour Press
Printed and Bound in the USA by Courier Westford, Inc.

British Library Cataloguing in Publication
data available

Library of Congress Cataloging in Publication
data available

ISBN 1–85230–961–X

Contents

Picture credits viii

1 The Sacred Berlitz Stone 1

2 Textual Confusion 18

Original Texts? 19

No End of Contradictions 20

A Gift from Heaven 22

Eve and the UFO 26

Nothing but Legends? 28

Heavenly Disputes 31

Frankenstein's Zoo 33

Light for the Ark 35

The Business of the Flood 37

Another Point of View 39

The Sex-Hungry Angels 42

Science and Theology 44

The Right Selection 46

Making Selections 50

Enoch Once More 52

An Eye-witness Report 53

When Angels Mutiny 54

A Rather Troubled Ascension 55

3 The Return of the Gods 61

The Apocalypse 62

Prophets of Our Times 64

Believers and Unbelievers 66

Was Jesus the Messiah? 67

The Messiah of Islam 70

Praised Be the Stars! 72

The Golden Age 74

Star Wars 76

Ancient Science 78

Impossible Dates 79

Karma *Remains Eternal* 82

Waiting for the Super-Buddha 85

Psychological Cover-Ups 87

Seeds From Heaven 88

Gods of Yesterday – Gods of Tomorrow 91

Who Will Return? 93

Goodbye Papa! 94

Exegesis Through the Ages 97

Upsetting Old Values 99

The Seeds Bear Fruit 101

Return in Other Forms 104

4 Tracking Down the Truth 109

The Message of the Gene 111

Machines to Make Us Transparent 113

Not of This World 116

Jurassic Sparrow 118

Artificial Intelligence 120

Not Right in the Head? 122

Right in the Head After All? 126

The 'Ringed' People 128

A Trojan Horse 129
Hybrids of the Future 133
Falsely Programmed? 134
SETI Without Europe 136
Submitting to Censorship 140

5 The Great Deception: Conspiracy of Silence and the Latest Research 144
The Opener of the Ways 148
The Sensational News Concealed 149
Knowledge of the Ancients 151
Dating the Sphinx 153
Discrediting Gantenbrink 156
Scholarly Error 157
Squandering Trust 162

Index 166

Picture Credits

Acknowledgement is made to the following for permission to reproduce their illustrations:

Erich von Däniken, 1, 2, 5, 7, 8, 9, 10, 11, 12, 13, 14, 15, 17, 18, 19, 22, 23, 28; Rudolf Eckhardt, 3, 4; Rudolf Gantenbrink, 20, 21, 25, 26, 31; Martin Lüthi, 30; NASA, Washington, DC, 6; Bernd Poser, 16; Bruno Senger, 29; Phil Payter Graphics, 24.

1

The Sacred Berlitz Stone

Dear Reader, before I launch into the real theme of my book, here is a short but rather tall story that – as I hope will become clear – has some relevance to my argument.

The scene is the future, after some immense catastrophe in which the world as we know it has gone under. In their efforts to understand past epochs of civilization – through such surviving relics as a simple Berlitz translating computer – the survivors' descendants develop an inevitably falsifying mythology and religion; like all religious beliefs, it is built up around a core of truth, but it is so overlaid with false assumptions and interpolations based on their own experience and ignorance that the obvious, simple truth is increasingly shrouded in mystery.

In the abbey of the Sacred Berlitz, children were accepted as novices at the age of 15. This particular year there were just eight boys and ten girls present at the ceremony. The abbot spoke with concern about the small number in their 'birth-year'. Most of them had grown up within the abbey precincts; their parents worked there, serving the Sacred Berlitz. Besides the lay brothers and sisters, there were also berry-gatherers, hunters and craftsmen of all kinds, as well as midwives and healers. They were all united in the wonderful task of giving birth to as many babies as possible, and bringing them up strong and healthy. Since the Great Devastation, the only human communities in the area were few and far between; the abbot suspected that their ancestors might have been the only ones who had survived.

No one, not even the erudite abbot himself and his Council of Learning, knew what had happened at the Great Devastation. Some thought that the people of those times had possessed terrible weapons and mutually annihilated each other. But there was not much support for this view. It was hard to imagine that such dire weapons could exist. Also, tradition said that those early people had been happy and had enjoyed great abundance and prosperity. Why then should they fight each other? It was illogical. A likelier possibility, aired in the Council of Learning, was that some mysterious infection had decimated humanity. But this theory did not hold water either, for it contradicted the lore passed down in the first generations after the Great Devastation.

The three ancient fathers and four ancient mothers who survived the Great Devastation had told their children that the catastrophe had suddenly broken over their heads one peaceful evening. These accounts were indisputable. They had been written down by the sons of the ancients in the holy *Book of the Patriarchs*. Every child in the abbey of the Sacred Berlitz knew the Song of Doom which the abbot sang each year on the Night of Remembrance. It was the only extant text from the ancient days:

> I, Gottfried Skaya, born on 12 July 1984 in Basel on the Rhine, had gone with my wife and my friends, Ulrich Dopatka and Johan Fiebag, as well as their wives and our daughter Silvia, on a climbing expedition in the mountains of the Bernese Oberland.
>
> Since it was already past six in the evening, we took a short cut on our descent from the Jungfrau mountain, using the tunnels of the Jungfrau railway. Because of building works at the top of the mountain, no more trains passed through to the valley at this time of day.

Suddenly the earth shook and portions of the granite roof hurtled down onto the rails. We were terrified, and Johan, the geologist, dragged us all into a rocky niche. We thought the horrific episode had passed when an immense thundering began. Under us the ground seemed to melt, we heard terrible rumblings, worse than in any storm. Thirty metres in front of us the lower tunnel wall collapsed. Then all fell silent again.

Johan thought that it was either a volcanic eruption – very unlikely in that area – or an earthquake. We had to make a steep ascent in order to reach the upper tunnel exit.

A few metres before the exit, the noise began. I have no words to express these rampagings of nature. First the wind hurled snow and lumps of ice past the tunnel opening, then followed trees, cliffs, and whole roofs of hotels from below in the valley. There were crackings and explosions such as human ears had never heard before. The wind howled and raged, shrieked and bellowed; everything flew through the air, was whipped up 1,000 metres into the sky, then hurled around. The earth shuddered, the elements screamed. Cliff walls of granite cracked open like cardboard boxes. It was only because we were in the shaft of a tunnel, whose lower opening was filled with debris, that we were protected from the appalling storm. God Almighty be praised!

The terrifying winds continued for 37 hours. We had no strength left; we lay huddled together in apathy, our arms intertwined, in our refuge. We just wanted the mountain to collapse upon us. No one can imagine what we suffered.

Then came the water. Amidst the howling and racket of the winds we suddenly heard a rushing thunder. It was like a torrent and cascade of endless oceans. Gigantic water-fountains seethed and gurgled, hissed and pelted upon the cliff walls. Like a storm battering the sea-coast, new wave-mountains kept rearing their huge heads, collapsing over each other, thundering down into the valley, forming immense whirlpools which sucked all life into the depths. It seemed that all the waters of the earth had mingled in a mighty confluence. We wanted to die, and screamed our terror from bursting lungs.

For eight hours the water boomed; then the winds sank, the groans of nature quietened, and all became still. Battered by this torture, speechless with pain, we gazed into each others' eyes. At last Johan crawled on all fours to the small opening that was still left high up in the tunnel exit. I heard him uttering terrible sobs and

> struggled forward to his side. The sight that met my eyes dumbfounded me. My innermost feelings were ripped to shreds. Then I too began to cry bitterly; our world no longer existed.
>
> The peaks of all the mountains were flattened, as if planed by a giant file. There was no ice or snow anywhere, nor any green either. Wet walls of cliff shimmered in a bare, brown light. The sun could not be seen; and below in the valley, where the spa-town of Grindelwald had been, were now only the waves of an enormous lake.
>
> This took place in the year 2016 of the Christian calendar. We do not know if anyone else survived the Great Devastation. Nor do we know what happened. May God Almighty stand by us!

The eight youths and ten girls listened awe-struck to the Song of Doom. The abbot, Ulrich III, had delivered it in a powerfully sonorous voice. After a short meditative pause he turned to the novices and said: 'Now enter the Hall of Remembrance. Examine with reverence the relics of the ancient fathers. You have been chosen, together with your brothers and sisters, to honour and understand these relics.'

Full of expectancy, the young novices entered the long, dark wooden building, which until then they had known only from its exterior. Lay sisters had lit wax candles, and the relics of the ancients shimmered in the flickering light. There were the shoes of the holy ones – Gottfried Skaya, Ulrich Dopatka and Johan Fiebag. The shoes of their wives were not there. The shoes were made of a strange material that felt soft like leather but was not leather. Not even the members of the Council of Learning knew what it was. A lay brother patiently explained that there might have been animals with such skin in the ancient times, which had been destroyed in the Great Devastation.

Christian, who was 17 years old and the eldest novice, slowly lifted his hand. 'Dear brother,' he humbly asked, 'what is the meaning of the writing upon the shoes of the holy Johan?'

With a good-natured smile, the man replied: 'All that we can decipher are the letters REE at the beginning, and the letter K at the end. We have not yet been able to ascertain its meaning.'

Once more Christian lifted up his hand: 'Dear brother, were there animals upon whose skin grew writing in the ancient times?'

'You're a bright fellow,' replied the lay brother in a rather annoyed tone. 'All things are possible for God the Almighty.'

In a grotto of the darkened room lay the survival pouches of the ancient fathers. The lay brother patiently explained that these were described in the *Book of the Patriarchs* as 'rucksacks'. The word 'sack' meant pouch; but the first syllable 'ruck' was not understood, nor the connection between the two parts of the word.

The novices were once more confronted with a riddle: the survival pouches were made of various coloured cloths, which were actually not cloth at all. Like the shoes of the holy Johan, these pouches felt soft and flexible; yet in the 236 years of the New Age, they had not disintegrated. The novices, in their joy, praised Almighty God: such a wonderful world they lived in, full of mysteries.

Another of the relics was the glistening rope that had been found in the survival pouch of the holy Ulrich Dopatka. No one knew what the strange, elastic, yet untearable material was of which the rope was made. But in the holy *Book of the Patriarchs* it was written that this material was called 'synthetic' – a word from the ancient times, obviously, whose meaning not even the erudite brothers of the Council of Learning understood.

The novices experienced strange feelings as the lay brother showed them a scrap of 'wrapping paper'. It was the same dull, shining brown as that on which the holy Gottfried Skaya had written the Song of Doom. How they must have suffered, those worthy, holy ancient fathers! What wonderful knowledge and materials they must have had in the ancient days!

The first viewing of the relics lasted for an hour. The novices saw unfamiliar tools, mysterious pencils and objects which were named 'clocks' in the holy *Book of the Patriarchs*, including a partly transparent clock with only one hand, which always pointed towards the setting sun. The lay brother gave a demonstration: whichever way he turned the clock, the hand swung back immediately towards the place where the sun set.

The initiation ceremony reached its culmination. The novices were looking forward with feverish excitement to the moment when they might for the first time catch a glimpse of the Sacred

Berlitz Stone. Accompanied by the swelling chorales of the lay brothers and sisters, they stepped into the innermost sanctum. In all grottos and on all ledges oil-lamps were burning; the air was rich with the heavy perfume of pine-oil. Before them, in the ceiling of the hall, was a circular hole through which shone a sunbeam, illuminating the altar. And there, upon a small stool, rested the Sacred Berlitz Stone, the greatest treasure which the abbey possessed.

Abbot Ulrich III gave a prayer of thanks. Those present listened with deep emotion and bowed heads. The formal part of the festival of initiation ended with the words: 'Holy Berlitz, we thank you for this gift from the heavens!' All the novices now gathered round their abbot. Carefully he lifted the Sacred Berlitz Stone from the stool and held it out towards the young people with a smile of radiant joy.

The stone was roughly as big as a hand. It was black with many small buttons, upon which, if one looked closer, individual letters could be discerned. The upper part of the stone contained a slit, under which was a dully glowing, grey background. Next to this, in a clear script, were the letters 'BERLITZ'; and below, in smaller letters, the word 'Interpreter 2'.

With his finger-tip, Abbot Ulrich III pressed the buttons with the letters of the word 'LOVE' on. Immediately the letters 'L-O-V-E' appeared on the grey background. It was eerie; the novices hardly dared breathe. Then Ulrich pressed another button, and directly under the letters 'L-O-V-E' appeared, as though written by a ghostly hand, the letters '*A-M-O-U-R*'.

'Halleluja!' cried Ulrich and lifted his gaze up to the beams of light pouring down through the roof.

'Halleluja!' rejoiced the novices and the brothers and sisters of the choir.

'The power of the stone is preserved! Praised be the Holy Berlitz and its enduring power!'

Once more the abbot pressed the buttons. This time the word 'H-O-L-Y' appeared; and shortly afterwards the letters '*S-A-C-R-É*'.

'Halleluja!' shouted the abbot to the roof, and 'Halleluja' echoed from the crowd. Ulrich III began, in faster and faster sequence, to press letters of other words on the Sacred Berlitz Stone. Each time, strange letters appeared beneath the words. It was a wonder not to be grasped by human understanding. The novices looked at one another with astonishment. They knew that they had been witnesses to a great wonder. It was a sublime moment.

At last Ulrich reluctantly and carefully placed the Sacred Berlitz Stone back upon the stool. With reverence and an expression of gravity he turned to the novices. 'The Sacred Berlitz Stone is a translating stone. With its help, the language of the holy ancient fathers can be transformed into other languages of the Old Age. The stone is holy, for it retains the eternal power of the sun. Three hours of sunlight suffices; then the stone will speak for twelve hours. Never has it disappointed the Council of Learning. It has helped us to understand the holy *Book of the Patriarchs*. It will help us, also, to decipher other writings of the ancient times, whose remnants are often discovered.'

Now Valentin, the second oldest novice, tentatively asked: 'Reverend Father Ulrich, where does the Sacred Berlitz Stone come from?'

'A wide-awake young man!' replied the abbot good-humouredly. 'Know, then, that the Sacred Berlitz Stone was discovered by the holy ancient father Ulrich Dopatka. In the *Book of the Patriarchs* is written how the holy Ulrich Dopatka found the stone. This happened two years, eleven months and nine days after the Great Devastation. The holy Ulrich Dopatka climbed the remains of that mountain which they called the Jungfrau. A few hundred metres below the peak, which had been destroyed in the Night of Destruction, there were ruins. In the *Book of the Patriarchs*, chapter 16, verse 38, it says even that these were the ruins of a scientific station which had once existed below the mountain's peak.'

The abbot paused for breath for a few moments before continuing: 'My young friend, the holy Ulrich Dopatka probably climbed the mountain which was called Jungfrau in the hope of finding something useful in those ruins. Perhaps he was guided by the spirit of the Holy Berlitz so that he would find the Sacred

Stone. The ways of God are many and mysterious!

'Tomorrow you will all begin to read the holy *Book of the Patriarchs*. In the coming years you will learn many things. Be obedient and humble. Praise God Almighty and the holy ancient fathers!'

In the *Book of the Patriarchs*, each chapter began with the words: 'My father told me ...' The original text of the book had been written by the sons of the first fathers – the patriarchs – and had consisted of 612 pages in total. Of the original text, however, only about a quarter remained. The script was very hard to decipher because it was so smudged and yellowed with age. Thank goodness the lay brothers and sisters had soon begun to make written copies.

The first eight pages, however, were different, for they had been written by the holy Gottfried Skaya on that 'wrapping paper' which the first fathers had had with them in their survival pouches. These pages were inscribed on both sides in a thin, black colour, whose composition no one understood. They bore dates of the old Christian calendar.

Then nothing more had been written down for many years, until the first writings on animal skins appeared. These had been written by the patriarchs, and the sons and nephews of the first fathers. They had introduced a new calendar, counting the years since the Great Devastation. The neat red letters of these documents shone upon the dark yellow background of the hides; often, several hides were bound together with plant-stems. Not until the year 116 after the Great Devastation did the descendants of the patriarchs begin to use chalk-paper: an underlay was made from woven plant-fibre, and upon this was smeared a thin layer of chalk. To make the whole thing smoother, the chalk was mixed with fruit-oils.

The novices took great pleasure in their studies. Their teachers were the older members of the monastery; any specific questions they had were answered by those who sat on the Council of Learning.

'Honoured Council-member,' asked a novice in the fourth week of study, 'why am I called Birgit and my neighbour here

Christian? Why is there a Valentin, a Marcus, a Will and a Gertrude? Where do these names come from?'

'These are the names which the first fathers gave their sons and daughters. There were three fathers: the holy Gottfried Skaya, the holy Ulrich Dopatka and the holy Johan Fiebag. They had four wives altogether, whose first names alone are known to us: Silvia, Gertrude, Elisabeth and Jacqueline. The first fathers procreated with these wives and brought forth children; in the first years after the Great Devastation, every wife gave birth to a child each year. All these descendants received names which the patriarchs knew from the ancient times. Does this answer your question?'

Then Valentin spoke: 'Yesterday we read Chapter 19; but we could not agree what was meant by the "Great Birds". Honoured Council-member, can you explain this to us?'

The honoured council-member hesitated for a moment, then smiled and stepped thoughtfully to the side wall, on which copies of the *Book of the Patriarchs* hung from rough wooden racks. He found the page containing Chapter 19, separated it from the others, and laying it before Valentin asked him to read out the text.

> Chapter 19, Verse 1: My father told me that his father Gottfried told him this parable as, one day at noon, a large bird flew over the valley.
>
> Verse 2: In my times, there were birds 200 times larger than that bird.
>
> Verse 3: In the stomachs of those birds sat men who feasted and drank.
>
> Verse 4: Through small skylights they could view the earth below them.
>
> Verse 5: These birds flew with rigid wings, faster than the wind, over the great waters.
>
> Verse 6: Beyond the great waters were houses so high that some of them touched the clouds. For that reason they were called 'sky-scrapers'.
>
> Verse 7: In the cities with the skyscrapers lived millions of people.
>
> Verse 8: We do not know what has happened to them. May God have mercy on their souls.

'Now, Valentin, what do you think this means?'

Valentin shrugged his shoulders. 'I don't really know. I can't imagine great birds in which people sit and even eat.'

'Do you doubt what stands written in the *Book of the Patriarchs*?'

Valentin was silent, but the alert Birgit spoke: 'The text comes from a Patriarch of the third generation after the Great Devastation. He emphasizes that his grandfather related this parable to his father. A parable must mean some sort of comparison.'

The novice Christian who sat next to Birgit and seldom contradicted her, because he loved her, interrupted in an unusually vehement way: 'I take the holy text at face value, even if I can't imagine giant birds in which people sit and eat. The holy Gottfried Skaya did not lie to his son – he was a living witness of the ancient times.'

The heated discussion which followed was interrupted by the honoured council-member: 'Enough, novices! The Council of Learning has discussed Chapter 19 on numerous occasions. We have also questioned the Sacred Berlitz Stone. The stone does not know other words for the great birds. Therefore they cannot have existed. The skyscrapers, it is true, are recognized by the holy stone. There must have been, then, great houses or towers, such as are described in the *Book of the Patriarchs*.

'Therefore it is our belief that the great birds in which people sat were a vision of the future vouchsafed to the holy Gottfried Skaya. You know of course that human beings cannot fly, yet wish that they could be like the birds in this respect. In accord with this wish, no doubt, the holy Gottfried Skaya envisaged a far-distant future in which people would fly over the water like great birds, without toil or exertion. It is likely that the young Patriarch made a mistake when writing down this account. He should not have put Verses 2–7 in the past, but in the future tense. In other words, not "There *were* birds which *were* 200 times bigger than that bird ...", but "There *will be* birds, which *will be* 200 times bigger." Do you understand, novices?'

All were silent. Marcus and Christian did not agree with the Council of Learning on this point. In his imagination, Christian was already conjuring great birds, made out of strong wooden beams, in which people sat and waved to those below.

From month to month the study of the texts became harder and harder. This was because much of the original material had been illegible and had therefore not been transferred to the excellent copies which had been made. In addition, there were many missing passages even in the original sources: gaps in the text, which made it hard to understand the whole. Most puzzling were the incomplete texts of the first generation – Chapter 3, for example, in which the cause of the Great Devastation was discussed.

> Verse 1: My father told me that his friend Johan, the geologist, thought it had been caused by a great meteor hitting the earth.
>
> Verse 2: The risk of being struck by a meteor or comet was to be statistically expected once every 10,000 years.
>
> Verse 3: The force of the collision ... [illegible] ... 20 times greater than the Hiroshima bomb.
>
> Verse 4: [The beginning missing in the original]...asteroids Geographos, Adonis, Hermes, Apollo and Icarus cross the earth's orbit.
>
> Verse 5: [The beginning missing in the original] ... a polar rift which led to a shift in the earth's axis.
>
> Verse 6: The north pole is now in the direction of sunset ... [illegible].
>
> Verse 7: What was once land is now under water; only the high mountains and high valley areas are not submerged.
>
> Verse 8: The mountains previously under the sea must now be exposed ... [the rest is missing].

Verse 1 was already difficult. The word 'geologist' was always mentioned in connection with the holy Johan Fiebag. But there was no explanation of the meaning of this word. The Sacred Berlitz Stone gave the word 'geology' – but what did it mean? The same was true of the incomprehensible words 'comet' and 'meteor'.

The honoured members of the Council of Learning were quite at a loss to explain the concept 'Hiroshima bomb'. They had examined this word in all its possible constituent parts, without being able to determine any meaning. 'Hir' could be read as 'here', 'Hiro' could, by making the 'i' into 'e', become 'hero'. And bomb, they had discovered with the help of the Sacred Berlitz Stone, meant something 'thrown' and 'exploded'.

The meaning of the middle part of 'Hiroshima bomb' was impossible to ascertain, although some council members believed that it referred to that distant land of the ancient times that was called 'China' at another place in the text. 'China' and 'shima' were similar. What did the whole word mean, then? Most likely 'what was thrown by a hero in China' or 'here exploded the hero from China'. This interpretation was disputed by other council-members, however, for it was known that only the three first fathers and four first mothers had survived the Great Devastation. Where, then, had the 'hero from China' come from?

The meaning of Chapter 4 was just as chaotic and difficult. Here, the son of the holy Ulrich Dopatka had written:

> Verse 1: My father told me that they had been very hungry in those days until they noticed that the waters were full of fish.
>
> Verse 2: In the first months they had still hoped that some aeroplane would appear.
>
> Verse 3: No aeroplane arrived though, but a UFO.
>
> Verse 4: They had been able to observe it for a long time, both men and women.
>
> Verse 5: The UFO had passed gently over the rocks at the lower shore.
>
> Verse 6: Some months later, the whole shore had begun to sprout and become green.
>
> Verse 7: Among the plants growing there, they had found many well-known crops: potatoes, maize, corn – in fact everything which people need for their nourishment.
>
> Verse 8: All had been very grateful and glad; but the extraterrestrials had not shown themselves again for many years, until the time when they came to find Gottfried Skaya.

The honoured members of the Council of Learning gave this chapter of the *Book of the Patriarchs* the title the Song of Hope. Verse 1 was clear, but Verse 2 contained an incomprehensible word: 'aeroplane'. The Sacred Stone gave only the word '*avion*', which the most erudite had associated with 'bird'. Through comparison with three other places in the text, it was known that 'aero' meant 'to do with the air'. But what was the meaning of 'plane'? The Sacred Berlitz Stone indicated that this meant something flat. Whichever way they turned it around, they could not make sense of it: 'flat-bird', 'air-bird', 'air-flat', 'flat-air bird'. It was not hard to agree with an older council member, who asserted that this must contain a small error: that the son of the holy Ulrich Dopatka must have written an 'l' in the wrong place by mistake. It should not say 'aeroplane' but 'aeropanel' – an old word, perhaps, for a wall or protection against the air or wind. It had no doubt been cold and windy in the first months after the Great Devastation. That is why the Patriarchs had hoped for something to give them protection from the cold wind, but it had obviously not been forthcoming. This interpretation was compelling and generally accepted by most.

Yet the difficulties in interpreting the rest of Chapter 4 remained insuperable. What did the Patriarchs mean by 'UFO'? It must have been something which they could spend a long time watching. Somehow this UFO had something to do with the crops which started sprouting by the shore. The UFO must surely signify Almighty God, for the crops had all been destroyed at the time of the Great Devastation. And now, thanks to the UFO, they had reappeared. This must, then, refer to God's bounty and eternal goodness, which preserved the first fathers and mothers from starvation. That is why they were all – as Verse 8 so wonderfully described – very glad and full of gratitude.

But what about the word 'extraterrestrial'? Whatever this was, it had come, later on, to find the holy Gottfried Skaya once more.

The members of the Council of Learning knew the word 'terrestrial'. It meant 'bound to the earth'. 'Extraterrestrial' must therefore mean something that came from beyond the earth, that was clearly not bound to it. This must, then, refer to Almighty God or one of His messengers. There was no doubt about this in the Council of Learning. The Almighty God must have chosen Gottfried Skaya to be the one to whom He sent one or more messengers. The phrase in Verse 8 allowed no other possible interpretation: '... but the extraterrestrials had not shown themselves again for many years, until the time when they came to find Gottfried Skaya.'

The highly intelligent and perceptive monks could not do other than search for the meaning of these things. The answer came like a flash of illumination. Almighty God had allowed the whole world to be destroyed, so the Great Devastation must have been a punishment which the Lord brought upon mankind – a purification of the earth. But because the Almighty God in His endless goodness had not wanted to destroy humanity utterly, he had chosen a small band of pure people to survive the destruction. These were to found a new race of men.

These ideas were confirmed when the perceptive thinkers of the monastery managed to work out the meaning of the name Gottfried Skaya. 'Skaya' was interpreted as 'sky' or 'heaven'; and the Sacred Berlitz Stone responded to '*Gott*' with 'God', and to '*fried*' with 'peace'. It was clear, therefore, that 'Gottfried Skaya' represented the new peace which God had made with mankind, after He had purified the world through the Great Devastation.

Brother Johan, a descendant of the holy Johan Fiebag, to whom this brilliant interpretation occurred, was awarded the Order of Thinkers for it.

After four and a half years, only 3 of the original 18 novices had remained faithful to their studies. The others worked in the abbey or in the fields; and all the female novices had, without exception, given birth to their first children.

Marcus and Valentin were mostly in complete agreement with the prevailing ideas and opinions, and held inspiring lectures in the abbey. Christian remained doubtful and sceptical. He had tried many times to gain access to the Revelation of the Holy Gottfried Skaya. But only the abbot himself was allowed to see this. Christian's intelligent perspicacity did not like to be fobbed off with mysteries and faithful acceptance, so he decided to become abbot himself.

The path to the top, to becoming abbot, was a long and arduous one, frequently paved with intrigue of all kinds; one had to perform a balancing act between the Council of Learning and the chief officials of the area outside the abbey. Christian's task was also made more difficult by the fact that he could never reveal the real truth about his motives, or share his innermost thoughts with anyone.

As the years passed, Christian became an ever lonelier figure. He spent much time shut away with his studies, isolating himself. The people around him believed this was due to the inner fire and dedication that burned within him. They were right, but did not know that this fire was fuelled by his doubts about the interpretation of the texts. Christian wanted to know, not to believe. Textual study had been rendered an impossible tangle of learned commentaries. Each council-member believed his own thoughts to be most apt and tried to impose his personal view. In the newer copies of the *Book of the Patriarchs*, larger and larger portions of text had been left out altogether because, according to the learned council-members, 'they had no meaning and only confused the issue'.

In Chapter 45 of the *Book of the Patriarchs*, it was written that only a few days after the Great Devastation, wood had come floating on the waters and the first birds had reappeared; and after a few weeks, green shoots and sprouts had begun to show themselves in holes and cracks in the rocks.

The Council of Learning took this to be a miracle dispensed by the hand of God. Christian disagreed. Various birds might well have escaped the Great Devastation by creeping into caverns in the rock. Pollen and seeds could have been whirled up into the air, and later sunk to earth and begun to grow. The same might be true of the various kinds of small animals which little by little reappeared. They might have crawled into all sorts of places to shelter from the Great Devastation.

The endless debates about all this were very wearing. For example, in the original text was written (Chapter 32, Verse 6): 'Thank God Uli's lighter still works; we were able to fry the fish …' But in the newer version this was altered to: 'God sent Ulrich Dopatka a fire, so that the first parents might warm their food.' That was a falsification of the text! In spite of his vehement expressions of disagreement, and the luke-warm support of Valentin and Marcus, Christian was outnumbered. The Council approved the new version.

Just as absurd was the debate about Chapter 44, which had come to be called the Period of the Angels. The original read as follows:

> Verse 1: My father told me that people of the Old Age had journeyed through space.
>
> Verse 2: Several expeditions had been sent to the moon, and returned safely to earth again.
>
> Verse 3: The technology involved had been very costly, therefore different nations had co-operated with each other by sending their scientific advisers to work on these projects.
>
> Verse 4: For the year 2017, one year after the Great Devastation took place, a second expedition to Mars had been planned.
>
> Verse 5: To avoid tensions and dispute, all nations involved with these projects had been kept informed of their technological progress.
>
> Verse 6: Exchange of information had taken place through scientific messengers and advisers.

From the *Book of Astronomical Facts* (Chapters 49–51), it was known that 'moon' referred to the small night light; and that Mars was the neighbouring outer planet of earth. The names of all planets were known, as well as the structure of the solar system.

In spite of the clarity of this information, the Council of Learning refused to accept the concept of space travel. Now one of the words which the Sacred Berlitz Stone gave in response to 'messenger' was the word '*ange*', from which the Council derived 'angel'. Obviously these messengers had been angels, there was no doubt about it; this was confirmed by the fact that at nine other places in the text the word angel fitted and made perfect sense.

The new version of Chapter 44, with the addition of extremely enlightening commentaries, now read:

> My father told me that in the Old Age people had observed the

> heavens. They dreamed of travelling safely to the moon and returning safely. In those times, the angels visited the various nations. They warned human beings of the Great Devastation and that it was wrong to worship the planet Mars. In order to avoid tensions and dispute, all nations were informed of these warnings. The angels themselves spread this information.

According to Christian's ideas, these alterations falsified the original text; yet it was approved by the Council of Learning. The Council, it was now said, was 'inspired by the spirit' and therefore had the authority to adapt the incomprehensible texts into a reasonable, accessible form.

Christian was 49 years old when he was elected to the position of abbot. In honour of the holy Gottfried Skaya, he gave himself the name Abbot Gottfried II.

2

Textual Confusion

'Those who cannot attack the thought, instead attack the thinker.'
(Paul Valéry, 1871–1945)

The texts written millennia ago, which have been passed down to us, contain a wealth of idiocy; they are a teeming melting-pot of fantasy – part myth, part legend – some of which, also, are considered to be holy books. Many of these far-fetched stories lay claim to absolute truth. Their original textual sources are supposed to have been personally dictated by God, or at the very least by some archangel or other, or heavenly spirits; or perhaps by a saint or 'inspired' person in the gnostic sense. ('Gnosis' is understood, nowadays, as an esoterically influenced philosophy, world-view or religion. But the word 'gnosis' derives from the Greek and means 'knowledge'.)

It is indisputable that these texts contain much humbug and wishful thinking. Revered leaders are elevated and glorified; daydreamers transform cloud-shapes into signs from heaven; everyday occurrences like death are described as journeys to the underworld. Still worse, our ancestors, in their thirst for knowledge, and moved by their true belief and desire to understand, falsified and obscured the texts. Events which no doubt had nothing to do with each other in the original versions were connected. To 'clarify' things, additions were made which suddenly – hey presto! – were passed on as original sources. Morality, ethics, belief and tribal history became interwoven; foreign elements from other cultural traditions were mixed in;

and texts were cobbled together whose source and original meaning will now, no doubt, never be deciphered.

This mish-mash is understandable. We are talking about texts which are thousands of years old, and about our ancestors' uninterrupted efforts to make sense of them. The state of muddle in the ancient texts is brought home to us when we realize the degree of confusion that can be caused in a much shorter time than millennia.

Take one example: every faithful Christian is convinced that the Bible is, and contains, the word of God. And as far as the Gospels are concerned, there is widespread belief that Jesus' companions had more or less written down his utterances and prophesies as they occurred. People think that the evangelists first experienced the wanderings and miracles of their master's life, then shortly afterwards noted down what had happened. This 'chronicle' of Jesus' life is endowed with the appellation 'original text'.

Original Texts?

But in fact – and every theologian with a few years of study under his belt knows it – this is all quite false. Those 'original texts' that provide such a rich seam for theological sophistry do not actually exist. What do we actually have? Copies which, without exception, were made between the 4th and 10th centuries after Christ. And these copies – about 1,500 in number – are taken from earlier copies; and not one copy is exactly the same as another. More than 80,000 discrepancies have been counted. There is not a single page of these so-called 'original texts' which does not contain contradictions. From copy to copy, verses were altered by authors who thought they knew what was meant and could express it in a way better suited to the needs of their time.

These biblical 'original texts' teem with thousands upon thousands of mistakes that are not hard to expose. The best-known, the *Codex Sinaiticus* – which, like the *Codex Vaticanus,* dates from the 4th century AD – was found in 1844 in the Sinai monastery. It contains no less than 16,000 corrections, by at least seven different hands. In several places the text has been altered several times and replaced with a new 'original text'. Professor Dr Friedrich Delitzsch, a specialist of the highest ability, found about 3,000 copy-mistakes in this text alone.[1]

This all becomes understandable when we realize that none of the evangelists was actually a contemporary of Jesus, and no contemporary wrote down an eye-witness account. Not until the destruction of Jerusalem by the Roman emperor Titus (AD 39–81) in the year 70, did anyone begin to write anything about Jesus and his team. The evangelist Mark, the first of the New Testament, penned his version at least 40 years after his master's crucifixion. Even the Church fathers of the first centuries after Christ agreed – on this, if on nothing else – that the original texts had been tampered with. They spoke quite openly of 'additions, desecrations, erasures, improvements, and wholesale destructions' of texts. In this regard, the Zurich specialist Dr Robert Kehl wrote:

> It has frequently happened that the same passage has been first corrected by one hand, then 'corrected back' again by someone else to give a quite opposite meaning, according to which dogma was currently the fashion of a particular school of thought. In any event, individual corrections – and general, systematic corrections even more so – produced a totally indecipherable chaos.[2]

Anyone who possesses a Bible can check up on this blunt conclusion. Just a few examples are sufficient: compare, for instance, the Matthew and Luke Gospels with Mark. The former two assert that Jesus was born in Bethlehem. Mark says he was born in Nazareth.[3]

No End of Contradictions

It would be nice if the theologians, at least, could agree about something! Instead they take up their conflicting positions, vehemently defending their own corner, sometimes just annoyed, sometimes elevating themselves to righteous anger in defence of their interpretations. For the layman it is quite impossible to force a way through the undergrowth of contradiction and distortion. But it seems to me that the theologians themselves, in spite of their hotline to God, are continually barking up the wrong tree.

If even the texts from periods we are well informed about – after all, we know something about Roman history – are so distorted and adulterated, what can we expect of texts which are several thousand years old? These ancient texts, no matter from what geographical or religious origin, are a hotch-potch, a mixed salad. One can drown in the thousands of pages of commentary that dedicated researchers of integrity and linguistic knowledge have written. The only thing they do not do is agree with each other, even within a single generation, let alone over longer periods.

It is my conviction that this salad of commentary on the ancient texts of humanity – even though sharp minds have poured on it a much-praised dressing of scientific research, analysis and comparison – has not advanced our knowledge by one iota. Centuries of thought and deep philosophizing by unarguably great and learned minds have not produced any certain answers, let alone any proof of the existence of God, the gods, the angels or the heavenly hosts. The literature of exegesis, of interpretation of religious texts, fills whole libraries, but no one can make head or tail of it any longer. The results obtained accord, at best, with the opinion of a particular school of thought, and change over time, according to the 'flavour of the month'. Not that it matters: each new generation neither knows nor cares what its predecessors thought.

In his dialogue *Phaedrus*, Plato cites Socrates as follows:

> At Naukratis in Egypt it is said that there dwelt one of the old gods, the same one in whose name the bird called Ibis is sacred. The god's name, though, was Theuth. He it was who first ordained the numbers and their harmonies, then the art of measurement and the star-lore, also board- and dice-games, as well as the letters …
>
> This god Theuth gave writing to the pharaoh of those times, with the words: 'This art, O King, will make the Egyptians wiser and of better memory; for it has been invented to aid remembrance and understanding.'
>
> The pharaoh did not agree, and contradicted the god Theuth:

> 'This invention will make striving souls more forgetful ... They will come to rely upon the outward signs of this writing; therefore they will no longer have inner and direct remembrance. Only outer memory will be aided by your invention, not inner remembrance.'[4]

He was right. The 1000-year-old scripts can only tell us of something that happened – perhaps – at some time and in some form or other. They cannot help us know what did happen.

Who knows, God – whoever that may be – may well have created other worlds long before this one. In *Jewish Tales of the Ancient Times*, one can read:

> The Lord created a thousand worlds in the beginning; then he created still more worlds; and they are all as nothing compared with him. The Lord created worlds and destroyed them, he planted trees and tore them up by the roots, for they grew haphazardly and each one got in the way of the other. And he continued to create worlds and destroy them, until he created our world. Then he spoke: 'This one pleases me; the others I did not like.'[5]

A Gift from Heaven

Was it really a human being who, in a lengthy phase of developing intellect, suddenly got the idea of scribbling down meaningful signs? Of course! Of course? Ancient traditions tell us that written script was invented 2,000 years before the world's creation. Since there was obviously no parchment available at that time, nor cattle-hide, nor metal, nor wood, this book existed, we are told, in the form of a sapphire stone. An angel by the name of 'Raziel, the same who sat by the river which flowed from Eden', gave this strange book to our first ancestor, Adam. It must have been something special, for it contained not only everything worth knowing, but also foretold all that would happen in the future. The angel Raziel assured Adam that he would find in it everything which 'will happen unto thee until the day that thou diest'.

Not only Adam was to benefit from this miraculous book, but also his descendants:

> Thy children also, who will come after thee, down to the very last of the race, will know from this book what will happen month upon month, and what will happen between day and night; to each one will be known ... whether misfortune or hunger will afflict him, whether corn will be plenty or scarce, whether there will be rain or drought.

A dictionary, or even a whole encyclopedia, is nothing compared with such a super-book! The authors of such a work must be sought amongst the heavenly host, for after the angel Raziel had given it to Adam, and even read to him from it, something astonishing happened.

> And at the hour that Adam received the book, a fire went up upon the bank of the river, and the angel rose in the flames up to heaven. Then Adam knew that the messenger was an angel of God, and that the book was sent him from the holy King. And he preserved it in holiness and purity.

Even specific details of the curious book's contents are noted. The inventiveness of its authors, who lived in some grey dawn of time, is hardly to be surpassed:

> In the book were engraved the higher signs of holy wisdom, and two-and-seventy kinds of knowledge were contained therein, which were divided into 670 signs of the highest mysteries. Also 1,500 keys, such as are not entrusted to the holy ones of the higher world, were concealed within the book.

Old Father Adam read the book with great diligence, for it alone gave him the power to name every object and animal. But when he transgressed, the book 'flew out of his hands'. Hocus-pocus.

Adam cried bitterly and walked up to his neck into the waters of a river. As his body became bloated and spongy, the Lord had mercy. He commanded the archangel Raphael to descend to Adam and return the wonderful sapphire stone. But it does not seem to have helped mankind very much.

Adam left the magic book to his ten-year-old son Seth, who must have been a particularly attentive young fellow. Adam told him not only about the 'strength of the book', but also 'in what its power and wonder consisted. He also spoke to him of how he himself had used the book and that he had hidden it in a cleft in the rock'. Finally Seth received instructions for using it, and for 'conversing with the book'. He might only approach it in reverence and humility. Moreover, he should not eat onion or garlic or other spices before using it, and should wash himself thoroughly before he did so. Adam hammered it into his son that he should never approach the book in a frivolous frame of mind.

Seth kept to his father's instructions, learnt from the holy sapphire stone throughout his life, and finally constructed '... a golden chest, placed the book therein and hid the chest in a cave in the town of Enoch'.

There it stayed until 'it was revealed in a dream to the patriarch Enoch the place wherein the book of Adam lay'. Enoch, the cleverest man of his time, did not delay; he made his way to the cave and waited. 'This he did in such a way that the people of that place should notice nothing.' Through some kind of parapsychology or other gnostic means, it was revealed to him how he should use the book. And 'at the very moment in which the meaning of the book became clear to him, a light dawned upon him'.

It must, rather, have been a whole chandelier, for Enoch

> ... knew now all the ways of the seasons, of the planets, of the lights which each month fulfil their services; also he knew the name of each cycle and orbit, and knew the angels who steer their courses.

Wonderful! The story is not as easy to untangle as it seems, however: it is not simply to be found on two consecutive pages of *Jewish Tales of the Ancient Times*. There are many small continuations and additions, fragments in many different, separate passages. I have not embellished the story by a single word, just tried to thread the pearls, as it were, onto one chain. So what happened to the book?

With the help of the angel Raphael, it reached the hands of Noah. Raphael explained to him how the book was to be used. The book was still 'written upon a sapphire stone', and Noah, who refounded humanity after the flood, learnt to understand, with its help, all the paths of the planets, also 'the paths of Aldebaran, Orion, Sirius'. He also learnt from it '… the names of all the different spheres of heaven … and the names of the heavenly servants'.

It is not quite clear to me why Noah was so interested in the paths of Aldebaran, Orion and Sirius; nor what use to him it was to know the names of the 'heavenly servants'. After the flood, I would have thought the survivors might have had concerns of quite another sort. Oh yes, and Noah laid the book 'in a golden shrine, and brought it at the very first into the ark'.

> And also when that Noah came out from the ark, the book went with him all the days of his life. In the hour of his death he gave it unto Sem. Sem gave it unto Abraham. Abraham gave it unto Isaac, Isaac gave it unto Jacob, Jacob gave it unto Levi, Levi gave it unto Kehat, Kehat gave it unto Amrom, Amrom gave it unto Moses, Moses gave it unto Joshua, Joshua gave it unto the elders, the elders gave it unto the prophets, the prophets gave it unto the wise men; from generation to generation was it passed down unto King Solomon. To him also was the book of mysteries revealed, and he became wise beyond measure … He raised great buildings, and through the wisdom of the holy book did everything prosper which he undertook … Happy is he whose eyes have seen, whose ears have heard, whose heart has understood the wisdom of this book.

This far-fetched story about Adam's book could be slotted into the 'fantasy' section without further ado if it were not for little details which make one begin to wonder. I can understand the desire to confer such a book upon Adam – for our lonely ancestor must have got his knowledge from somewhere although a book would not, strictly speaking, have been necessary. Adam was, no doubt, a fairly intelligent sort of chap, who learnt what he needed from daily experience. I can also understand that once a book had been introduced into the story, the chroniclers wondered where it had gone to, and so started making up its descent through the generations.

What does not quite fit with all this, however, is the idea of the sapphire stone. Whoever first came up with this story could only have imagined books made of paper, parchment, clay, wood, or slate tablets, or perhaps animal hide, or writing carved on cave-walls. Where on earth did the idea of the sapphire stone come from? Even centuries ago, let alone millennia, the idea of a whole encyclopedia engraved on a precious stone was wholly incomprehensible. But not any more. In the computer age, dictionaries on microchip are perfectly possible. Scientists are also looking at the possibility of storing information in crystals. Now, according to the story, Adam had 'conversations' with this sapphire stone book. What?! What was the inventor of this story thinking of? And where did he get those specific details, the '72 kinds of knowledge', the '670 signs of the highest mysteries', and the '1,500 keys'? This is precise information which you do not just pull out of the air, let alone attribute to angelic gifts from above.

It is indisputable that people millennia ago were of a more believing disposition; but their belief was also deeper-rooted. As far as I know they may well have thought that any old brass was true gold; at any rate, their belief in the creation of the world remained unshakeable. Now angels were seen as something superhuman: they were the swords and messengers of the eternal God. You did not mess around with angels – they were to be feared. How, then, does a chronicler come to include an angel in his ancient science-fiction? The 'angel Raziel' brings Adam the sapphire stone book, Raziel being the same one who 'sat by the river which flowed from Eden'. A load of irreverent nonsense? As if that was not enough, the angel Raphael is employed to return the book to Adam again after the Fall.

I do not overestimate the capacity of this mysterious book; yet I have to ask why the author places such importance on certain star-constellations. Why do Adam and his descendants need to know the paths of Aldebaran, Sirius and Orion? There are simpler ways of making an earthly calendar.

Eve and the UFO

The angel Raziel, who brought the sapphire stone book, also 'rose in the flames up to heaven'; but not before 'a fire went up upon the bank of the river'. One can read about fire and flying chariots during Adam's times, in the apocryphal text *The Life of Adam and Eve*.[6] Although the extant version dates from AD 730, it is based on handwritten documents of an unknown age.

> Eve looked towards the heavens and saw a chariot of lights drawing near, pulled by four gleaming eagles, whose magnificent beauty no one born of mother's womb is able to express.

Ancient Mother Eve the first witness of a UFO? The same Lord who had created Adam and Eve, and from time to time walked at his pleasure in the Garden of Eden, also climbed aboard this UFO:

> And behold, the Lord of strength mounted upon the chariot; four winds drew it, the Cherubim guided the winds, and the angels of heaven went before ...

Adam also gleaned from the sapphire stone book the names of all the different spheres of heaven, as well as the names of the heavenly messengers. But what heavens are we talking about?

The *Jewish Tales of the Ancient Times* give us more precise information. The first sphere is called Vilon; from here mankind is observed. Above Vilon lies Rakia, where the stars and planets are to be found. Higher still is the sphere of Schechakim, and beyond this the heavens which are called Gebul, Makhon and Maon. The highest sphere of heaven, beyond Maon, is Araboth. Here the

> ... Seraphim dwell. Here are also the holy wheels and the Cherubim. Of fire and water are their bodies made. Yet they remain whole, for the water doth not extinguish the fire, nor doth the fire suck dry the water. And the angels give praise unto the Most Holy, blessed be His Name. But far from the glory of the Lord dwell the angels; they are distant from him a space of 36,000 miles, and do not see the place where His glory dwells.

The word 'miles' is, of course, not present in the original source, but rather an unknown measurement which was replaced by a translator with a term he understood. But the number 36,000 has not been altered. Yet a peculiarity of the story is that these different heavenly spheres are not only characterized by measurements of distance, but also of time. Between one heaven and the other are 'ladders', to cross which it takes periods of '500 years' journey'. If one looks at these accounts through modern spectacles, this is a distance of ten light years at a speed of 2 per cent of the speed of light.

All these stories and accounts go under the heading of 'tales and legends' – which are thoroughly unreliable – nothing but 'idiotic fables' as the theologist Dr Eisenmenger mocked over 200 years ago.[7] They are easy to dismiss. In contrast to 'history' they can be relegated to the realm of fiction; they are grotesque and wonderful, fascinating and outrageous. Such tales and legends of course completely ignore chronological time-sequence, and fail to have the slightest respect for historical fact. Legend is 'a people's speculation and fantasy';[8] yet it still remains a valuable link between historical research and science. Legend augments history; it attempts to fill the gaps and shed light on the darkness. Legend is not built on nothing; even if its points of view and interconnections do not agree with historical sources, it still remains the 'religious philosophy of a people's history'. Even the Greek geographer Strabo (about 63 BC–AD 26), who wrote the 17-volume *Geographica*, remarked dryly: 'It is not Homeric to tell stories without any grain of truth.'

Nothing but Legends?

Legends magnify what is great, weave magic around what is mysterious, embellish their heroes with imaginary powers. But legend is nevertheless not a web of lies. It always connects with actual historical personalities and real events. Often it tries to preserve what historians neglect or destroy. Every Swiss citizen, for instance, knows the legend of William Tell and the apple. The historians rejected and demystified it, but do the Swiss care? In some form or other, they are sure, the story must have taken place!

Legends are also, and have always been, international. (Elsewhere I have demonstrated the extraordinary connections between Bible stories and the traditional stories of the Central American Indians.)[10] The Jewish legends also doubtless contain easily demonstrable similarities with Persian, Arabic, Greek, Indian and even American traditions. They may have differently named characters and heroes, a variety of different gods and descriptions of natural phenomena, yet the core of the stories is closely related. Would anyone disagree that the legend of the flood can be found in various forms throughout the world?

In legends, all historical dates are disregarded. It does not matter *when* something happened, only that it did. This is true also of many holy books. Let us take as an example the biblical version of the Flood, with Noah and his Ark. People just had to *believe* in this story, until a sensational discovery was made on the hill of Kujundshik, the site of Nineveh. Archaeologists brought 12 clay tablets to light, which had once belonged to the library of the Assyrian king Assurbanipal. These relate the story of Gilgamesh, the king of Uruk, who was a mixture of man and god, and who set out to seek his earthly ancestor Utnapishtim.

To our astonishment, Utnapishtim gives us a precise description of the Flood; he recounts that the gods warned him of its approach and gave him the task of building a boat, in which he should seek refuge with his wives, children, relations and craftsmen of every sort. The descriptions of the tempest, the darkness, the rising flood-waters and the despair of the people who were left behind, still read today as a gripping, moving tale. We also read – as in the Bible – the story of the raven and the dove who were sent out; and how, at last, as the waters sank, the boat came to shore upon a mountain.

The parallels between the flood story in the *Epic of Gilgamesh* and in the Bible are clear – and undisputed by any researcher. What is fascinating in this similarity are the *differences*: there are different gods and different circumstances involved. The flood story in the Bible is recorded at second hand, while in the Gilgamesh epic the first person is used throughout, suggesting the eye-witness account of someone who actually experienced the Flood.

History and research books erase, wreck and destroy, but legends do not. They remain obstinately alive in folk consciousness, written continually anew after every episode of war and devastation. Legend is unfocused memory, the past's vague legacy to the future. That is why I stick to legend, trying to its old spirit with modern means.

If we survey the stories and traditions of humanity that have been handed down – and now I expressly mean all that exist on the face of the earth – it seems that some lord or other, a *highest, most holy one,* a *dear god,* created the first human being. He placed this being in the Garden of Eden, or in some other gloriously beautiful spot. According to ancient Jewish tradition, this Garden of Eden existed long before the world was created, already provided with all amenities:

> All its grounds and plantations, and also the sky-vault above it, as well as the ground beneath – all was there; and the earth and heavens were not created until 1,361 years, 3 hours and 2 blinks of the eye afterwards.

And people still wonder why the Garden of Eden was never found, in spite of determined searches! (I have documented this search, and its failure, in an earlier book.)[11] The experimental research station, Biosphere 1, with its Adam and Eve experiment, was most likely later recycled. And if I had ever been tempted to believe that our primal ancestors were the only two people in Eden, the Jewish legends tell me different: 'Serah, daughter of Asser, is one of the nine who came living into the Garden of Eden.' And who, we may ask, were the other six?

The 'highest' had decided to create the human being. Before he did this though, he went through the formality of asking his angel hierarchies what they thought of the idea. They were against it. 'The Lord stretched forth his finger and burned up every last one.' Once more the 'highest' asked the same question of other angels – with the same result. The third group of angels replied that the 'highest' would anyway do what he wanted, so he might as well get on with it. So he created Adam 'with his own hands'.

The first 'model' human was apparently superior in some respects to the angels. It particularly annoyed them to think that the human being would gain power over a whole planet, and might reproduce at will. Angels are, apparently, infertile and unable to reproduce. There was, therefore, jealousy in heaven.

Heavenly Disputes

> Ishmael was the greatest prince of angels in heaven; for all holy creatures and the Seraphim had only six pairs of wings each, whereas he possessed twelve pairs. And Ishmael went and united with all the highest armies of heaven against his Lord; he gathered his armies about him and descended with them, and began to seek a companion upon the earth.

Such a mutiny was not to be countenanced by the 'highest'. What had to happen, happened: the 'highest' cast down Ishmael and his army from the place of holiness. According to Jewish legend, the sin in the Garden of Eden had nothing to do with the famous apple, but with the fact that this ringleader Ishmael seduced Eve and got her pregnant. After the sexual act, 'she looked upon his face. And behold – he resembled not an earthly but a heavenly being.'

A crazy story? Completely unbelievable? Pure fantasy? Hardly. The stories which have been continually copied and reinterpreted through the millennia contain a common core – one which resurfaces amongst countless different peoples in widely separated parts of the globe: the temptation and seduction of the human being. What really happened in that far-distant, nebulous past? Let us remind ourselves: the whole Christian religion is based upon the idea that Jesus had to come to save mankind. Save it from what? From original sin. This happened in paradise, that wonderful Garden of Eden. Whether it was an apple or sex, the decisive event took place somewhere. Eve's seduction occurred through a snake or through an archangel cast out of heaven. Modern theologians, who feel very uneasy with the whole idea, have hit upon a solution: original sin never occurred. By saying this they are also pulling the carpet out from under the idea of salvation, but that is their problem really, not mine.

And now we are faced with a paradox: heaven is traditionally a place of unadulterated joy. Heaven is what people aim for after death. Everyone would like to get there, be free at last from worry, jealousy, misery and need. Heaven is the aim of all longing and dream, the fulfilment of all hopes. But hang on a minute! Something is not quite right. There was already a lot of jealousy, conflict and deadly warfare in heaven before human beings were ever created. Have we therefore misunderstood the concept of heaven somewhere along the line? Are the old texts speaking of a different heaven from the one in which Almighty God dwells?

The dilemma still remains, even if one wishes to reject or disregard the old Jewish traditions, or if one thinks one's own idea of heaven is superior. Eve's tempter was, whichever way you look at it, the cause of the original sin which altered everything. Even if this sin never occurred, it remains, in Christian belief, the reason for our later salvation through Jesus. Legend or not, if there was no original sin there is no logical need either for salvation. Whether the tempter was called Ishmael, Lucifer, or the Devil does not alter the fact.

As everyone knows from the Bible, God Almighty sent a flood to drown the human race. But why? He had previously made the primal human being 'with his own hands', and, as a timeless and eternal God, could foresee the future. He must have known in advance what was going to happen. Or did he perhaps not? Then the 'highest' would be something different from what I and millions of devout people imagine by God. The Jewish legends tell us that after Eve's seduction two separate races arose: that of Cain and that of Abel. The descendants of Cain behaved like animals:

> Exposed and naked went the race of Cain, both man and wife as the cattle of the fields. Naked went they around the market-place ... and men procreated with their mothers and with their daughters and with their brothers' wives openly in the street.

The guile and deceit of this race are described in the tales of Sodom and Gomorrah. The inhabitants of these cities adhered to neither law nor morality, doing just what they felt like.[12]

In addition to the general collapse of morality and sexual antics in Sodom, 'fallen angels' descended in crowds from the heavens and took 'human wives'. We cannot categorize this sort of angel as 'innocent'. Their progeny grew to be giants:

> From them sprang the giants, who were of mighty girth, and who stretched forth their hands to rob and plunder and to spill blood. The giants raised offspring and multiplied like creeping things: six of them at one time were delivered at birth.

This was evidently a pigsty of humanity, without any redeeming aspects – no chance then of separating good from bad. What could the 'highest' do but drown the whole brood and start all over again? Which shows us, however, that he cannot have been anything like the true God whom believers of all religions worship.

The 'fallen angels' are supposed to have produced giants. I have talked about these giants in several books and do not wish to repeat myself. Just briefly then: the *Jewish Tales of the Ancient Times* distinguish between different giant races. There were the Emites or Frighters, the Rephites or Gigantics, the Giborim or Mighties, the Samsunites or Sly Ones, the Avids or Wrong Ones, and finally the Nefilim or Spoilers.

That must have been a wonderful crowd that had gathered on the earth! In the apocryphal tales of the prophet Baruch,[13] they are even given an exact number: 'God sent the floodwaters upon the earth, and wiped out all flesh, and also the 4,090,000 giants.'

Where on earth, or off it, did the prophet Baruch get this number from? Of course biblical chronology is once again wrong from beginning to end in respect to the giants. David, who lived a long time after the flood, was supposed to have fought against giants with six fingers and toes, as the second book of Samuel reports (21:18–22) – chronological nonsense.

Frankenstein's Zoo

I am not astonished by the dates, which are a hopeless muddle, but by the *events*. The *Jewish Tales of the Ancient Times* tell of strange mixed-breed beings, curious life-forms which do not fit into any evolutionary sequence. There were ones who 'had only one eye in the middle of their forehead'; others who 'had the body of a horse but the head of a ram'; still others 'with a human head and the body of a lion'; finally, even human beings without necks, with eyes upon their backs, and – stranger still – 'beings with human faces and horses' feet'.

Is this absurd menagerie just an enormous joke, or the crazy delusion of a drunkard? Possibly. But I am bothered by the way such accounts are duplicated in various places. The Egyptian Manetho, for instance, tells of similar monsters. This Manetho was scribe and high priest of the holy temples in Egypt. The Greek historian Plutarch mentions him as a contemporary of the first Ptolemaic king (304–282 BC). Manetho lived in Sebennytos, a town in the Nile delta, and there wrote a three-volume work about the history of Egypt. He had been an eye-witness of the end of the 3,000-year reign of the pharaohs; he wrote his chronicle of the gods and kings as someone who knew the facts.

Manetho's original text has vanished, but the historians Julius Africanus (d AD 240) and Eusebius (d AD 339) took substantial passages from his work. Eusebius was Bishop of Caesarea, and an early Christian chronicler whose accounts became part of ecclesiastical history. Manetho asserted that it was the gods who had caused certain mixed-breed creatures and monsters of all kinds to arise. This is Eusebius' version:

> And they were said to have brought forth double-winged human beings; also others with four wings and two faces; and with one body and two heads, man and woman, male and female within one creature; still other human beings had thighs of goats and horns upon their heads; others had the feet of horses; others were horses behind and men in the front; there were also said to have been man-headed bulls and four-bodied dogs, whose tails emerged like fishes' tails from their backs; also horses with the heads of dogs; ... and other monsters, such as all kinds of dragon-like beings ... and a great number of wonderful creatures, variously formed and all different from one another, whose images they ranged one beside the other in the temple of Belos, and preserved.[14]

Manetho, via Eusebius, was certainly right about the images. Every good-sized museum today exhibits ancient sculptures of mixed beings. The Jewish and Egyptian legends are therefore not just pure waffle; they evidently tell of some former reality.

And if these monsters from Frankenstein's studio never existed, where did their inventors get the idea for them from? What brain nourished these strange creatures; and where did the masons and sculptors of ancient times find their models? No doubt from tradition, which is extraordinarily, painstakingly precise – almost annoyingly so – for a silly old legend.

The Bible tells, in the Book of Genesis, of the building of the Ark (6:15): 'The length of the ark shall be 300 cubits, the breadth of it 50 cubits and the height of it 30 cubits.'

The Jewish tales are still more precise:

> One hundred and fifty chambers shall be the length of its right side, 150 chambers shall also be the length of its left; 33 chambers shall be its breadth at the front, 33 chambers shall also be its breadth at the back. In the middle shall be ten rooms for cooking utensils, also five store-rooms at the left side; there are to be pipes to bring water, which can be opened and shut. The vessel shall be three storeys high; as the first level is, so also shall be the second and third storeys; in the lowest storey are to be housed the cattle and wild animals; in the middle storey shall nest the birds; the upper story is for men and the worm-creatures.

Light for the Ark

After the Ark had been daubed all over with pitch, so that every crack was sealed, it must have been very dark in the antediluvian vessel. But this was not the case apparently, for 'in the vessel hung a great pearl, which shone upon all the creatures with the power of its light'.

An astonishing aside, at this point. The *Book of Mormon* is the 'Bible' of the Church of Jesus Christ of the Latter Day Saints, a religious community which became big in the USA. This book was supposed to have been given to the founder of the Mormon Church, the prophet Joseph Smith (1805–44), by an angel. According to the Mormons, this book was preserved for millennia in the form of metal tablets hidden inside a hill. It was only by means of two translation stones, which Joseph Smith received from the angel Moroni, that he was able to translate the ancient script into English. The tablets tell the story of the

Jaredites, a people who left their old homeland at the time of the building of the Tower of Babylon, and crossed the seas to South America. Their ships were 'watertight as a barrel, and when the doors were shut, they too were as watertight as a barrel'.[15]

Yet it was not dark in their vessels, for the Lord gave to the Jaredites 16 shining stones, 2 for each ship, and these stones dispensed bright light during the crossing which lasted 344 days. It was probably the same mysterious light-source as in Noah's Ark.

According to Jewish tradition, the Lord personally made Noah a drawing of the Ark: 'And the Lord drew with His finger before Noah, and spoke to him: "Behold, thus and thus should the Ark appear."'

The Mormons have something very similar. In the first book of Nephi (1:6) can be read: 'You should build a ship in the way that I will show you, so that I may lead your people across the waters.'

Did the Mormons, then, copy their text straight out of some Jewish legend or other? Or did the Jews do this from the Sumerian Gilgamesh epic, or from the Babylonian epic *Enuma Elish*? The latter also contains a variation on the flood story, in which there is a surviving patriarch called Atra Haris and a god – Enki – who, as always, orders a water-tight ship without any openings. The light-source and the compass are there as well.

The question of who copied what from whom cannot be answered. We do not have to assume plagiarism just because there are similar details in these legends and holy books. What gives us the right to exclude the possibility that the original text of the *Book of Mormon* was in fact engraved on primeval metal tablets? Surely it is only our Christian-Jewish vanity that makes us reject such an idea. And the fact that the flood story is known in slightly different forms in other cultures also does not prove that the Jewish chroniclers must have stolen the idea. There would have been many descendants of the first generation after the flood, who all developed their own versions of the story.

The authors of these diverse legends lived in different lands, continents, cultures and religious contexts. No news was carried between these places; intercontinental travel was not yet common. And yet, from innumerable sources and from all corners of the globe come stories and traditions which are almost the same as each other. Did one and the same spirit dwell in all the different brains of these writers? Were they all plagued by identical thoughts? Never! Certain things cannot be invented. No power of imagination could, thousands of years ago, have been at work all round the globe in the same way at the same time. All these uniform accounts must derive from prehistoric events. Originally, accounts were given of an actual experience. During the millennia these have been embellished and adorned, each people attributing it to their own folk-heroes and prophets. But at the original core, the great event remained, around which all these legends crystallized.

The Business of the Flood

Which brings us to the second dilemma – original sin being the first. The holy books proclaim that our dear God caused the flood to punish the evils of humanity. This flood obviously did take place; scientific evidence has come to light which substantiates it.[16] In addition, an international team of scientists believes that it has located the remains of Noah's Ark near the peak of the mountain Al Judi, the very mountain in the Ararat region upon which, according to the Koran, Noah's Ark was meant to have landed. The leader of the expedition, the geophysicist David Fasold, explained to journalists that they had used ground-radar to obtain excellent pictures. These images were so clear that one could even count the planks in the walls of the hull. And Professor Salih Bayraktutan, the director of the Geological Institute of the Ataturk University of Ankara, told journalists from the *Observer*: 'This is a structure built by human hands, which can only be Noah's Ark.'[17]

Did our loving God really order the building of the Ark? Whoever he is or was, this mysterious figure knew what he was doing, for he wanted to save at least some people from devastation. So he gave, either to one or to several people, according to different traditions, directions for building a ship. He even made plans and drawings with his own hands and/or dictated the exact dimensions. He dispensed mysterious, shining pearls or stones, and even compasses. Then the Great Devastation began.

Why so complicated? If God – and again I mean the God of all religions – wanted to get rid of some misguided angels or giants or evil people, he could surely achieve this with one symbolic wink of his eye. Or, as the Koran, the holy book of the Muslims, asserts: 'When he decides upon something, he simply speaks: "Let it be" – and it is.' (Sura 2, Verse 118). There is no need for a ship, plans, measurements, pitch or any mysterious light. All this ship-building business shows that someone or other either wanted things like that or could not do it any other way. Why technology instead of a miracle? The true God must have known that his involvement in the details of ship-building would, millennia later, only raise doubts about his omnipotence. Being omniscient, he would also know that there would one day be innumerable different reports and accounts of the flood. So why ship-building instead of a clearly divine solution? Miracles are known to be insusceptible to the calculations of critical reason. So what kind of a God was it who caused the flood and yet helped with plans and measurements for the Ark?

But if he did *not* cause the flood – if, in other words, he had nothing to do with the wholesale drowning of humanity, if the flood was a natural catastrophe – then this God was not the one we know from religion. In this case humanity would have ascribed to a God the meting out of a punishment for which he was in fact not responsible. In which case belief is on a slippery slope. Whoever favours the natural catastrophe idea must, however, explain why flood stories are the theme of *international* legends, folklore and holy books.

And one more thing: the flood as natural phenomenon or cosmic catastrophe (caused perhaps by collision with a comet or meteor) does not alter the fact that the 'highest' had prior knowledge of what was going to occur. Otherwise he could not have warned his protégés, could not have directed the building of the Ark or dictated instructions to make it water-tight.

So far only one thing is clear: this god of tradition cannot be the true God whom all believers of all religions worship. So who is he really?

I assume that it is known that I believe extraterrestrials visited our earth thousands of years ago. I have written 20 books and made a 25-part television series on this theme.[18] I have also discussed at length the reasons for this visit and its technical details. I do not intend to go over that ground again at this point, nor to discuss again the countless archaeological indications supporting my theory that have been found all over the globe.

My concern in this book is with a 'palaeo-seti' philosophy (palaeo = old, ancient, from the Greek *palaios*; seti = search for extraterrestrial intelligence), with a theory and edifice of ideas that illumines the sense or nonsense in religious views and beliefs, and opens a new path of thinking about these matters. My intention is certainly not to found a new religion nor, as my critics maliciously assert, a 'substitute religion'. Religion demands faith – which has no place in my investigations. Religions offer promises, even beyond death – I am not promising a thing. Religions build churches and temples where they worship their gods and holy ones – apostles, saints and prophets. In palaeo-seti philosophy there are neither temples nor worship. Religions also ultimately demand adherence to certain ethical norms – there is not the least trace of that in my supporters or me. And finally, religions insist on some yearly financial tribute. Do you, dear Reader, feel financially exploited by buying this book or borrowing it?

Another Point of View

As the giant mother-spaceship of the extraterrestrials cruised into our solar system, the ETs aboard had already known for a long time about the third planet. Only upon this blue planet did all the conditions for life exist. The strangers discovered a wealth of all forms of life, amongst which were our primitive ancestors. Though dumb and slow-witted, these represented the highest form of life upon the earth at that time. The aliens therefore took one of the creatures and altered it genetically – no longer, these days, such an unthinkable idea.

At some point or other, a group of extraterrestrials found that their experiment with the first *Homo sapiens* had succeeded, and that they could leave the earth to this indigenous human being. He was certainly cleverer than all the other crawling and flying creatures; he also had the ideal tools to undertake whatever he wanted – his hands. For this being to multiply, a female was needed – Eve, or whatever the name of our primal mother may have been.

The first intelligent human beings had no speech – how could they have done? Their direct ancestors were apes, they grunted and roared. So the extraterrestrials decided on a training programme. The *Homo sapiens* pair were placed in a protected garden – Biosphere 1 – and taught speech, as the book of Genesis (11:1) reports: 'And the whole earth was of one language and of one speech.' Adam was able, ultimately, to give a name to everything! The programme would also have included moral education and practical instructions for developing agriculture and crafts.

Another group of ETs experimented with the earth's animals. Why should they do that? A space crew in a gigantic spaceship, a so-called space-habitat, would certainly know other solar systems and planets besides the earth. At the very least their own solar system must have been familiar to them. Many of these other planets may have been larger or smaller than our planet, may have lain closer to or further from their respective suns; may therefore have been cooler, drier or wetter, and subject either to stronger or weaker gravity.

We know that there are myriads of life-forms on the earth that have adapted to the most inhospitable climates and conditions. The polar bear sleeps on the ice, something I would not recommend to a lion; a kangaroo takes giant leaps, while the tortoise creeps along; certain kinds of snake have adapted to tropical climates, and freeze in the cold. Surely it would be an obvious thing to want to experiment with the genetic material available on the earth, to find out which animals are best fitted for certain environmental conditions, and also which are most resistant and best at surviving. Is that an absurd idea?

We ourselves have done, and do, the same. Not – until now – by genetic means, but through breeding. We have bred German and Swiss cows which now graze happily in the tropical climate of Kenya; we have combined different breeds of oxen to produce stronger cows with a higher milk yield; we have crossed goats with sheep ('geep'); we have cross-bred varieties of grain to make them better adapted to a new environment; and now we have started producing genetically engineered vegetables. There is absolutely no knowing what our scientists may still come up with; who is to say that we will not suddenly genetically engineer a person who lives to the age of 240?

This is how the monsters and mixed-breed beings appeared that had not previously existed on earth. The human beings talked excitedly about them; these 'divine' creatures astonished and frightened them. And once they had died out or been destroyed in the flood, these horror-movie beasts walked straight into folk memory. They were elevated to myth and legend, symbols of a far-distant time when the gods had created all kinds of beings.

I am not, however, underestimating the human power of imagination. The Greek poet Homer (about 800 BC) described the sirens in the adventures of Odysseus, who sang so bewitchingly that seafarers lost their will and memory. Although Homer does not describe these sirens in any detail, the imagination of later authors depicted them as winged women with the feet of birds. There was also the Greek Hesiod (about 700 BC), who made up the monstrous Medusa from whose head snakes writhed and flickered, and whose appearance was so terrifying that it turned people to stone. Of course Hesiod never actually saw a Medusa. We also know the legends of the flying horse Pegasus and the phoenix which rises from the ashes. All these and much more arose from human imagination, upon which all folk tales depend. But imagination does not stem from nothing; it needs points of reference to spark it off. Even if our logical reason still resists the idea of a zoo of monsters that is supposed to have lived once upon a time, this resistance does not alter two unavoidable facts:

- Ancient writers and historians described these creatures and also asserted that they had been created by the gods.
- The stonemasons and stuccoists of thousands of years ago preserved these mixed beings for eternity.

The Sex-Hungry Angels

In the mother spacecraft, meanwhile, mutiny had broken out. Some of the higher-ranking officers had quite different ideas from the commander, the 'highest'. Whether the leader of the rebels was called Ishmael, Lucifer or anything else has little relevance. The legend described him as 'the greatest prince among the others'. In the science-fiction story *Star Trek* he would no doubt be called First Officer. Whatever his name, Ishmael alias XY seems to have had more power than the rest of the crew, for he was the only one to possess '12 pairs of wings'. Ishmael and his renegades lost the battle on board and were thrown out of 'heaven'. This does not seem, to begin with at any rate, to have bothered them much. They probably thought that their technical know-how would ensure that they regained the upper hand.

Hardly had they arrived on earth than the expelled crowd developed a hearty appetite for sex. In the legend, the leader Ishmael promptly seduced Eve: 'And behold, he resembled not earthly but heavenly ones.' Other crew-members snapped up, according to taste, pretty girls and also boys. Even steadfast believers in the Bible cannot get past this passage from the book of Genesis (6:1):

> And it came to pass, when men began to multiply on the face of the earth, and daughters were born unto them, that the sons of God saw the daughters of men that they were fair; and they took them wives of all which they chose.

The learned dispute which has raged since time immemorial around that little phrase 'sons of God', and which has produced thousands of pages of conflicting and contradictory commentary, will raise only a bored smile from anyone with a bit of inside knowledge. 'Sons of God' is variously translated as 'giants', 'children of God', 'fallen angels', or perhaps 'renegade spirit beings'. It is enough to make one want to scream! A single little phrase turns belief on its head! Every specialist who has a sure knowledge of Hebrew can tell you exactly what these syllables mean: 'The ones who had descended were similar to men and much larger than human beings.'[19] But one is not allowed to say what one thinks. I do, though, and without any ifs or buts.

The old objection that extraterrestrials could never pair up with earth-people has long since been dismissed; I do not need to repeat myself on this point. ('And the gods created men in their image ...')

In this drama of prehistory, the 'highest', the commander of the spaceship, obviously possessed better maps than his renegade crew. With concern he watched the goings-on on earth. The cross-breeding of ETs with people gave rise to creatures which were wholly inconsistent with the planned race of *Homo sapiens*. This was the *original sin* of mythology. Human beings were now inheriting the wrong genetic messages. 'And it repented the Lord that he had made man on the earth, and it grieved him at his heart,' it says in Genesis (6:6). The 'highest' must somehow now interrupt the 'human being' experiment and begin afresh. But how? The renegade angels probably possessed powerful weapons, they could hide away in caves and hold out inside buildings. There was no possibility of hunting the evil ones individually.

We cannot gather from the legends and religious texts whether the flood was caused intentionally, or whether a large meteor collided with the earth. An artificial flood is possible – we still do such things today – and meteorites are constantly raining down upon the earth. Whichever it was, the 'highest' must have been informed about the exact time that the flood would take place – which is how he could warn the good ones, and advise them about their ship-construction.

Science and Theology

I myself do not see much future for the kind of theology we have known until now. Theologians may believe in revelations, but they will never manage to make rational what is, by its nature, irrational. That does not mean that I dispute the scientific approach of systematic theology, in which texts are compared with known historical events, hand-written scripts are examined, and an attempt is made to evaluate a number of different accounts through comparative analysis. For example, which prophets, and when, made reference to the Messiah? Which of their statements are incomprehensible, and to which should we attach less importance? What do all these statements have in common, and with which other account does a particular prophet's description agree? If theologians would simply call this 'science', I would have no argument with them – apart from the term 'theology' itself. This derives from *theos* (god) and *logos* (word), and thus means 'the word of God'. But this is precisely what theology is *not*. Of course it is true that all theologians are convinced that they are concerned with the 'word of God', otherwise they would never have chosen this profession. But such a conviction is already *faith*. They have *faith* that the holy and not so holy texts once came from the mouth of God, that he dictated them or revealed them to the chosen few. But what remains of these texts once the ingredient of faith is withdrawn?

What is left are the texts themselves. They have simply lost their holiness. They may remain venerable because of their great age. We can treat them with respect because they describe occurrences from a time inaccessible to history. We can analyse them scientifically because they contain much that is of great interest. Once belief in the holiness of these texts is dispensed with, we can really start getting to grips with them. It is actually our idea of their holiness that prevents a modern analysis of their significance.

On the other hand, palaeo-seti philosophy is also just a point of view, a theory; it provides a very helpful foundation, but cannot yet be proved. Is theology any different? Are there any precise, *scientific* proofs for its assumptions? It is well known that there is nothing more subjective than taste or opinion – therefore there is no point in arguing about them. Yet people do argue, because the generation gap coupled with the spirit of the times causes uproar in their inner lives. Some want to hold fast to the safe stronghold of belief; others want explanations that hold scientific water. 'Science' comes from the word *scientia* – knowledge.

The 'knowledge' of theology is useless to exact science. It is full of contradictions and ultimately remains a matter of belief and feeling. The same is true of palaeo-seti philosophy. But the latter develops a clear thread, a train of thought which employs reason and makes the incomprehensible more accessible. Palaeo-seti philosophy makes sense of what was previously nonsense. Occultists might as well put away their crystal balls, the members of secret brotherhoods can close up their shops; for the wares of belief that sold so well through millennia are less and less in demand. Only modern scientific knowledge can provide us with an understandable interpretation of the past. And this is not by pure chance either; it lies in the nature of things. Apples fall when they are ripe. And my grandfather could never have conceived of the ideas which I now propound. Space travel was unheard of in his day, he knew nothing at all of genes and genetic engineering; and angels were for him inviolable messengers of God. He would have thought a hologram was a vision and a television speaking glass. Praised be the Sacred Berlitz Stone!

It is not because we are drawing close to the end of the millennium that the veils have lifted, but because science and technology have flung wide the gates. If people had not started discussing the possibility of space travel or invented the computer or unlocked the secrets of the genetic code until the year 2100, we would not, until then, have been in a position to examine the questions which such things have thrown up. Let us imagine that my great-great-grandfather had stumbled upon some wonderful find 200 years ago. Say he had discovered engraved tablets which, when deciphered by learned men, told of a journey from a far-distant world to the earth, and of the travellers' unfriendly reception by earth's inhabitants. What would people have made of such a text 200 years ago? An enormous power of imagination would have been attributed to the unknown author; the text would have been seen in terms of allegory and symbolism. People would have extracted from it some kind of moral, such as that we should be friendly to strangers even if we do not know where they have come from. But the real possibility of space travel would have been quite beyond their grasp.

So I believe that we should bring a modern view to bear upon the age-old questions of humanity. Such questions may well have become much easier to solve than in the days of my great-great-grandfather. We do not suffer from excommunication or witch-hunts any more; and modern means of communication allow the rapid development and spread of new theories. I can understand why some people, entrenched in their old beliefs, want to fight a rearguard action to dam the flood-tide of new discovery. They may hold it up for a little while, but no power on earth is going to stop the arrival of the future. Things which are forbidden by religion and ideology in one country have a habit of turning up in a still more radical form in another.

Critics continually ask me what makes me so sure that I am on the right track. They tell me my views are nothing more than a fixed idea, and unprovable. They also accuse me of using only those passages from legends and mythology, in a very selective way, which support my theories.

The Right Selection

But why am I accused of doing what everyone must do, given the enormous wealth of material? Every book I read is a selection which the author has chosen to support his views. The objection that scientific investigations do not approach the material in this way is a pure fantasy in which only unkissed students believe. In the last four years I have devoured about 300 theological works – and the conclusion of each one supported the author's opinion. Countless cross-references are used, especially in PhD theses, to show that the author's opponents are misguided in one or other respect. The flood of literature on every subject has become so immense that no author in the world can still retain an overview, or take account of all the works of his predecessors. One *has* to select, silently throwing the ballast overboard in the process. The subject-specialist has a broad knowledge of the opinions he opposes, which the layman does not particularly want to bother with; the publishers and booksellers still less. We should recognize that selection is unavoidable and admit honestly that an author says what he wants to say and makes clear the line of enquiry he is pursuing.

Religious texts are riddled with morality and ethics, which do not interest me in the least. That is why I do not bother with the hundreds of pages of the prophets' warnings, threats, prophecies and instructions. It is not my business to explain to the reader why one should not eat pork, and on what grounds one should repudiate one's wife. Every specialist knows, anyway, that prophets' utterances are very rarely authentic or original. Later generations have added, extended and spiced up these texts to their taste. And also, in respect of religious chronology, what good are passages such as 'Tarah begat Abraham, Abraham begat Isaac' when Abraham probably never existed?

What? But there are texts about Abraham, stories have been written about him; and in the apocalypse of Abraham, experiences are described in specific detail. That is right; there are such texts, and ones which are very useful for my work. But this does not prove that we are dealing here with original sources from Abraham's hand, or from those close to him. In the *Chronicles of Jerahmeel*,[20] which are based on still older sources, it is asserted that Abraham was a very great astrologer and magician. He was said to have received his knowledge directly from the angels. We, as people of a Christian culture, have had it drummed into us that Abraham was the progenitor of humanity; but in fact the researchers are not even agreed that he actually existed, and what his name signifies.

Franz M. Böhl, Professor at the University of Leiden, states:

> The name Ab-ram, which appears only in Genesis 11:26–17:5, means 'the sublime father' or 'the father is sublime'. One can take

> the word 'patriarch' itself as a translation of this name ... Abr-aham is probably only a dialect variant, an expansion of the more common name Abram.'[21]

This passage was written in 1930, but later investigators came to a similar conclusion. Five years after Professor Böhl, the *Journal of Biblical Literature* noted succinctly: 'Abraham was originally not a personal name but the name of a divinity.'[22]

The 60 years of Abraham research which have passed since then have shed no new light on the subject. In a publication from Yale University, I read the following noteworthy passage: 'We will probably never be in a position to prove that Abraham really existed.'[23]

What need is there then, in the face of this theological confusion, to take heed in my work of the chronological dates of any prophets' sayings? Especially since the same sort of doubt hangs over other prophets as well. Ezekiel, one of the prime witnesses in my case for palaeo-seti philosophy,[24] had to pass through countless transformations throughout the centuries. In a work which appeared in 1981, no less than 270 treatises on the prophet are examined.[25] Two hundred and seventy wise heads devoted years of their lives to Ezekiel research. The figure of this prophet underwent, in the process, extraordinary transformations. Originally his word was beyond reproach; then he became a 'visionary'; subsequently a 'dreamer' and 'idealist'; and most recently he has been considered a 'cataleptic' – that is, a schizophrenic subject to fits. The Ezekiel texts were also dissected. Experts on semantics discovered that style and vocabulary showed that they were written by more than one author. The poor prophet was declared to be a 'pseudo-Ezekiel', whose book had been cobbled together 200 years after the death of Christ from a variety of other different texts.[26]

A hundred years ago, however, the theology professor Rudolf Smend could still write:

> There can be no doubt that the text is based on a visionary experience, which is by no means just attributable to a particular convention of written style.[27]

And today? Most theologians believe that the book of Ezekiel is the work of several editors, which includes the work of the prophet himself as well as additions inserted at various periods.

Who can really hold it against me, then, if I select the freshest leaves from this salad of confusion? It is a salad which also contains indigestible spices. There are names and dates which appear in the holy books, which belong in the salad as little as sliced shoe-soles. Take this passage from the book of Genesis, for instance (15:13 and 16):

> And he said unto Abram, Know of a surety that thy seed shall be a stranger in a land that is not theirs, and shall serve them; and they shall afflict them *400 years* ... but in the *fourth* generation they shall come hither again ...

The British archaeologist Kathleen M Kenyon noted sourly to this:

> The chronology contradicts itself. To accept that their stay lasted 400 years, and yet simultaneously to be told that the fourth generation after the entry to Egypt would be involved in the Exodus, are two so obviously incompatible assertions that one is forced to regard them as unhistorical.[28]

Theological viewpoints are not only opaque, they also alter from one professor, and from one decade, to another. So what are we left with? The mysterious accounts themselves. The sort of stories in which the author writes in the first person, that is, retells a personal experience. As in legend and myth, religious literature retains a core, a nugget of truth. It is that mysterious aspect that later editors hardly altered. Why not? Partly because they did not understand it; the mystery stuck to the prophets' words and was handed down to succeeding generations. Partly also because they did not dare to put their own words into a venerable prophet's mouth in a wholesale fashion; they would then have had to lie in the first person. The original, personal experience of 'I saw ... I heard ... the highest spoke to me ...' came from a primeval, ancient source. The later editors only tinkered around with it, trying to make sense of the incomprehensible. And because they did not understand it themselves, we are left today with perfect chaos. If only they had just copied the old texts without making alterations! But for a thinking person this is well-nigh impossible. We cannot even do it nowadays. We already have versions of the New Testament in the form of comics, and other still worse adulterations, ostensibly to make the holy text relevant to our times. Yet, 'from unclean means comes an unclean result' (Mahatma Gandhi, 1869–1948).

Making Selections

My process of selection overlooks all that is completely incomprehensible to present-day understanding. That does not mean that we will not be analysing these things again from a different perspective in 20 years' time. Whoever says that such a process is unscientific, that one should not proceed in such a way, should take a look at the Jewish scholars who have faced exactly the same problem for centuries and millennia. They did not understand the significance of the old texts either; so every word, every phrase, was turned this way and that, and constantly reinterpreted and reformulated. There is written proof of this in the many *midrash* books. The well-known *midrashim* literature contains the textual research of the wisest Jewish minds over many centuries.[29]

Such interpretation goes on for hundreds of pages. New names, new points of view. And all of it only proves that the greatest Jewish scholars no longer understood the original texts.

So how do I make my selections? How do I proceed? How can I know better than the scholars of the past and decide which passages were once original and which were not?

When the life of Abraham[30] is described in terms such as that angels descended at his birth, and that he held his own against King Nimrod of Babylon, I assume that this is all the pious addition of later editors. They were concerned to give Abraham a strong profile and a fittingly glorious origin. But whenever Abraham – or whoever he was, the name is irrelevant – starts speaking in the first person, I prick up my ears. I home in on such passages, especially when they describe some astonishing episode connected with space, which later editors could not have invented, because they would not have had access to such detailed knowledge.

In the text which theologians call the *Apocalypse of Abraham*, the author – let us call him XY – describes two heavenly beings who descend to earth.[31] These two heavenly beings carried Abraham up to the heights, for the 'highest' wished to converse with him. Abraham relates that they were not human and that he was very afraid of them. He describes them as having a shining body 'like a sapphire'; they took him upwards in smoke and fire, 'as if with the force of many winds'. Once arrived in the heights, he saw a 'glorious light beyond description' and great figures who shouted words to each other 'which I do not understand'. And for anyone who still has not grasped where he has got to, he makes it even clearer: 'But I wished to fall back down to the earth; the high place where we stood was at one moment upright, at another moment it turned upon its head.'

So someone is telling us – in the first-person, narrative form – that he wanted to 'fall back down to the earth'. It is logical to assume therefore that he was *above* the earth. And why could a later editor not have invented the text? Because no one could have known that gigantic spacecraft, like the space stations of the future, always turn on their own axis. It is only through the centrifugal force caused by a spaceship's own rotation that an artificial gravity can be engineered. And what does Abraham's *Apocalypse* say? 'The high place where we stood was at one moment upright, at another moment it stood upon its head.' Coincidence? Just foolish fantasy? Why does Abraham insist that these beings were not human and that their suits glittered like sapphire?

Such texts are actually crystal clear. And their time has come. Modern man has had about enough of having religious fairy-tales forced down his throat. There is a new, modern interpretation of the old texts and traditions which makes all clear in an instant.

Just before I launch into a quite new chapter, I would like – for the last time – to kindle an old fire that I have often returned to in the course of the years. There is hardly a single one of my books which does not mention the prophet Enoch. I shall not go back exhaustively over the old ground now; but I would still like to hammer a few posts into it as markers that modern exegetes will find hard to ignore.

Enoch Once More

Who was Enoch? The ancient Jewish tales describe him as 'a king over men', who ruled for '243 years'. He was full of wisdom, which he dispensed to all the world.

He was the builder of the great Egyptian pyramids, according to the geographer and historian Taki al-Makrizi (1364–1442). In his work *Hitat*, he mentions that Enoch was known by four different names: Saurid, Hermes, Idris and Enoch. The following passage is quoted from *Hitat*, Chapter 33:

> The first Hermes, known as the threefold one in his qualities of prophet, king and wise man ... read in the stars that the flood would come. Then he ordained that the pyramids should be built; and in them he hid treasures, texts and scripts, and everything which might otherwise be lost, so that they might be preserved.[32]

The Arabic word *idris* means 'progenitor' or 'primal father of wisdom'; and for both Jewish and Christian theology Enoch is the seventh of the ten original patriarchs before the flood. Enoch was the father of Methuselah, who was supposed to have lived to the biblical age of 969.

In the Old Testament, Enoch shows up for just five verses (Genesis 5:21–4). And then it says: 'And Enoch walked with God: and he was not; for God took him.' And hey presto he vanished! In Hebrew the word *enoch* means 'the initiate' or 'the insightful one'. Thank God, this initiate made sure that his knowledge did not vanish without trace – to the annoyance of the orthodox, who would prefer it if it had vanished into thin air – for he was an assiduous writer. And this is what started all the trouble.

There are two books, which are not included in the Old Testament, but are among the apocryphal texts. The Church Fathers who put together the Bible did not know what to do with the Enoch texts. They excluded them because they did not understand them. But the Ethiopian Church ignored the orders of the ruling ecclesiasts; so the Enoch book ended up in the Abyssinian canon. A Slavic variant of the same book also turned up. Textual comparisons between the two by specialists showed conclusively that they both derived from one original source, written by one author – none other than Enoch. So who was he?

I am continually astonished by the shortsightedness of different exegetes. If a text fits in with their convictions, they consider it genuine. If it does not it must be false. The Book of Enoch is not only written in the first person, but the author also continually makes reference to his own authorship – as though he was afraid that future minds might be too narrow to recognize the fact. I would like to cite two textual examples which demonstrate clear indications of Enoch's authorship.

An Eye-witness Report

> In the first month of my 365th year, on the first day of the first month, I, Enoch, was alone in my house ... and there appeared unto me two great figures of men, such as I had never seen before upon the earth ...[33]

> This is the full and true teaching of wisdom, written by the author Enoch ... and now my son Methuselah, I tell you all and write it down for you. I have revealed all these things to you and passed on to you the books which concern them. Preserve, my son Methuselah, these books from your father's hand, and pass them down to the future generations of the world.[34]

You cannot get much clearer than that. The original source of the Book of Enoch comes from the Enoch who lived before the flood – for he refers to his son as Methuselah. To assert that that is all just a pre-Christian falsification is to accuse the author of unadulterated lies. To attribute the Book of Enoch to sources other than the Enoch who lived before the flood is a disgrace to the discipline of textual research. It is also a horrendous example of manipulation of the devout, who are supposed to swallow whatever predigested fare is offered them. Of course researchers also attempt to dismiss the embarrassing Enoch texts as 'visions'. This little word is stretched to cover everything that passes our understanding. The 'vision' proponents overlook the fact that Enoch expressly says he was in a state of wakefulness. In addition he gives his family exact instructions for the period of his absence. This also cannot have been a 'death-bed vision', for after his conversations with the 'angels' he returns fit as a faithful fiddle to his relatives. Not until much later does he vanish into the clouds in a fiery chariot.

So what is so important about this Enoch book? Quite simply, it represents the corroboration of palaeo-seti philosophy. As in the Old Testament, Enoch gives an account of what happens when angels mutiny.

When Angels Mutiny

In the Book of Enoch (6:1–6) it says:

> As the children of men multiplied, lovely and loving daughters were born unto them. When the angels, the sons of heaven, caught sight of them, they lusted after them and they spoke to one another: 'Let us take to ourselves wives from amongst the children of men, that they bear us children.' Then their leader Semiaza spoke to them: 'I fear that you will not go through with this thing; then would I be made to bear the penalty for a great transgression.' Then all replied unto him: 'Then let us swear an oath and bind ourselves not to give up this plan but to carry it out.' So all swore an oath and bound themselves to it. There were in all 200, who in the days of Jared descended from the peak of the mount of Hermon.[35]

If this is not mutiny of the 'sons of heaven', what is it? It is quite clear what was going on, for (7:1–6):

> All of them took wives unto themselves. Then they began to go to them and to do impure acts with them. And they taught them the arts of magic and herbs, and showed them plant-lore. Then their wives were with child and gave birth to giants, 300 feet tall. These devoured all the provisions of the other people. But when there was no more to feed them with, the giants turned against them and ate them up. And they began to devour birds, wild animals, creeping things and fish; and also the flesh and blood of one another. Then the earth cried aloud against these fiends.

The pre-flood scenario is described in realistic detail, even if it appears unbelievable to us nowadays. The good angels – the ones who had not become involved in the mutiny – observed all this from above. They reported to the 'highest', and he decided on action: 'The whole earth shall go under; a flood of water shall come upon the whole earth and destroy all things.'

What is remarkable about the Enoch book are the many details not to be found in any other text. In Chapter 69, Enoch even lists the names of the leaders of the mutiny and describes all their ranks and functions!

So what happened to Enoch? Where did his bones come to rest? Where is the temple or cathedral erected in his honour?

A Rather Troubled Ascension

It is not to be found on this earth. The Old Testament allows Enoch to disappear without trace. The Lord is supposed to have taken him. Or, according to which biblical version you follow, he rose up into the clouds on a fiery chariot. The ancient Jewish tales are more precise about his take-off.[36]

The angels, apparently, had promised to take Enoch with them, but the date of departure had not yet been fixed. 'I received word that I would journey into the heavens; yet upon which day I go from you I know not yet.' So Enoch gathered people around him and told them of everything that the angels had told him. In particular he told them not to hide his books away and keep them secret, but to make them accessible to future generations – which is a call I myself heed. After a few days of dispensing his wisdom, things became exciting.

> But it happened at the same time that the people were gathered around Enoch, and he was speaking unto them, that they lifted their eyes and saw the figure of a steed descending from heaven to the earth as though in a rushing storm. Then the people said unto Enoch what they saw, and Enoch spoke to them: 'For my sake has this steed descended to the earth. The time has come, and the day, upon which I go from you and see you nevermore.' Then also was the steed there, and all the children of men saw it with their own eyes.

It is clear that Enoch had been told by the heavenly ones that the take-off would be very dangerous for all bystanders. So he tried to hold them back. He warned the watchers several times not to follow him, 'so that you will not die'. A few hesitated and stayed well back, but the most obstinate wanted to get a good close look at Enoch's departure.

> They spoke unto him: 'We will accompany you to the place to which you go; only death itself will divide us from you.' Since they would not attend to his words, he spoke no more with them; and they followed him and did not turn back. And so it came to pass that Enoch rose up to heaven in a storm, on fiery steeds, in a fiery chariot.

This ascent to the clouds ended in death for all the observers. The next day, people came to search for those who had accompanied Enoch.

> And they sought them at the place whence Enoch rose up to heaven. And when they came to the place, they found the earth covered with snow, and upon the snow lay great stones like unto hailstones. Then each spoke to the other: 'Let us dig away the snow and see if we may not find those who accompanied Enoch.' And they dug it away and and found those who had gone with Enoch lying dead beneath the snow. They sought also for Enoch, but found him not, for he had risen into the heavens ... This took place in the year 113 of the life of Lamech, the son of Methuselah.

So we are faced with yet another impossibility – after the Fall and the flood. But we will by now have ceased to be astonished, for all previous textual interpretations are riddled with impossibility. We are supposed to believe that our dear loving God just stood back and watched as hundreds or perhaps thousands of observers burned to ashes, while their teacher Enoch rose to the heavens! What misdeeds had they done? They had listened to the wisdom of Enoch, they accompanied him to the take-off spot. Enoch rose in a storm, on a fiery chariot up to heaven, while below him the recipients of his wisdom burned, together with the earth and stones, to a snow-white ash. (Some forms of limestone become snow-white at great heat.)

None of these events – the Fall, the flood, the ascension of Enoch, or indeed the space travel of Abraham – fit with the image of a loving God. Why should an all-present God call Abraham to him to talk with him? He must know – being all-knowing – what Abraham is thinking and feeling. Why should our dear God need a spaceship that revolves upon its own axis above the earth? Why must God first send two figures to fetch Abraham? Why does he need 'fiery horses' to carry Enoch to heaven?

The answers to these questions are always the same: the 'highest', the God described here, can never in a thousand years be the same as the all-present Creator honoured by all religions (and by me). I consider it an insult to the true God to ascribe such mistakes and cruelty to Him. But if we replace God or the 'highest' with extraterrestrial space travellers, the paradoxical events become understandable. We can then understand who these fallen angels were, and why they satisfied their sexual lust. We can understand the reasons for a flood, and for the desire of the 'highest' to communicate with individual human beings; and we can understand why the many people who did not heed Enoch's warning were burned to death.

This also makes comprehensible people's fear of the day of judgement, of some kind of universal settling of accounts – for the 'highest' had promised to return ...

NOTES

1 Delitzsch, F, *Die grosse Täuschung*. Stuttgart/Berlin, 1921

2 Kehl, R, 'Die Religion des modernen Menschen', in *Stiftung für universelle Religion*, Vol 6a, Zurich

3 The Gospel of St Matthew begins with the line of descent of Jesus 'the son of David, the son of Abraham'. His ancestors are listed down to Jacob, who was Joseph's father. Joseph was Mary's husband. But what good is this line of descent if Jesus was not supposed to have been Joseph's offspring? (Jesus was thought, you will remember, to have been immaculately conceived.) Matthew lists 42 ancestors of Jesus; Luke, on the other hand, lists 76.

The evangelists are also in disagreement about the last words of Jesus on the cross. According to Mark (15:34) and Matthew (27:46), he called in a loud voice: 'My God, my God, why has Thou forsaken me?' According to Luke, on the other hand, he called: 'Father, into Thy hands I commend my spirit'. John's version is: '"It is finished": and he bowed his head, and gave up the ghost.'

Even the ascension – the most impressive event of the Jesus story – is reported in different ways. According to Matthew (28:16–17) Jesus commanded his disciples to gather on the mountain in Galilee. 'And when they saw him, they worshipped him: but some doubted.' Still doubted? Matthew has nothing more to add about the ascension.

Mark (16:19) has only a single sentence for the extraordinary event: 'So then after the Lord had spoken unto them, he was received up into heaven, and sat on the right hand of God.' It was that simple was it?

Luke (24:50–2) says that Jesus himself led the disciples 'out as far as to Bethany, and he lifted up his hands and ... while he blessed them, he was parted from them, and carried up to heaven.'

John, Jesus' dearest disciple, knows nothing about an ascension.

These are only a few examples of Bible texts which are accessible to everyone, which are differently translated from Bible to Bible, according to the wishful thinking of different Churches. (Extracts quoted here are from the King James version).

4 Plato, *Phaedrus*, Penguin, London, 1973

5 Berdyczewski, M J (Bin Gorion), *Die Sagen der Juden von der Urzeit*, Frankfurt am Main, 1913

6 Fuchs, C, 'The Life of Adam and Eve', in *Die Apokryphen und Pseudepigraphen des alten Testaments*, Vol 11, edited by E Kautzsch, Hildesheim, 1962

7 Eisenmenger, J, *Entdecktes Judentum*, Königsberg, 1711

8 Bergmann, J, *Die Legenden der Juden*, Berlin, 1919

9 Strabo, *Erdbeschreibung*, translated into German by Dr A Forbiger, Berlin

10 Däniken, E von, *Der Götter-Schock*, Munich, 1992

11 Däniken, E von, *Wir sind alle Kinder der Götter*, Munich, 1987

12 A little example, with some contemporary relevance, to illustrate this:

> The people of Sodom and Gomorrah set up beds in the street. Whoever entered their towns was seized and forced down upon a bed. If the stranger was shorter than the bed, three men pulled at his head, the others pulled at his feet. The man would cry out, but they took no notice, continuing to stretch him. But if the stranger was longer than the bed, three men stood either side of it and stretched him sideways until he was tortured to death. When the stranger cried out at this torment, they shouted back: 'This is what happens to one who comes to Sodom.'

13 Kautzsch, E, *Die Apokryphen und Pseudepigraphen des alten Testaments*, Vols 1 and 2, Tübingen, 1900
14 Karst, J, *Eusebius-Werke, Vol 5, Die Chronik*, Leipzig, 1911
15 *The Book of Mormon*, 16th edition, 1966
16 Tollmann, A and E, *Und die Sintflut gab es doch*, Munich, 1993
17 Bayraktutan, S, quoted in *Die Welt*, 17 January 1994
18 Däniken, E von, *Auf den Spuren der All-mächtigen*, broadcast between January and December 1993 on SAT-1. Also the books *Auf den Spuren der All-mächtigen* and *Raumfahrt im Altertum*, Munich, 1993
19 Agrest, Matest M, 'The historical evidence of Paleocontacts', in *Ancient Skies*, Vol 20, No 6, Highland Park, Illinois, 1994
20 Gaster, M, *The Chronicles of Jerahmeel*, New York, 1971
21 Böhl, F M Th, *Das Zeitalter Abrahams*, Leipzig, 1930
22 Albright, W F, 'The Names Shaddai and Abraham', in *Journal of Biblical Literature*, Vol LIV, 1935
23 Seters, J van, *Abraham in History and Tradition*, New Haven/London, 1975
24 Blumrich, J F, *Da tat sich der Himmel auf. Die Raumschiffe des Propheten Hesekiel und ihre Bestätigung durch modernste Technik*, Düsseldorf, 1973; Beier, H H, *Kronzeuge Hesekiel*, Munich, 1985
25 Lang, B, *Ezekiel. Der Prophet und das Buch*, Darmstadt, 1981
26 Torrey, C, *Pseudo-Ezekiel and the Original Prophecy*, New Haven, 1930
27 Smend, R, *Der Prophet Ezechiel*, Leipzig, 1880
28 Kenyon, K M, *Bible and Recent Archaeology*. British Museum Publications, London, 1987
29 *Midrash* is the work of interpretation, the searching for sense. I do not expect my readers to go out and buy the *midrashim*; I will therefore just take a few examples. The following comes from the *Midrash Bereshit Rabba*, which consists of more than a hundred chapters:

> God spake: Let us make man. With whom did God consult? According to Rabbi Joshua, in the name of Rabbi Levi: with the works of heaven and of earth. Like a king with two advisers, who did nothing without consulting them. According to Rabbi Samuel bar-Nachman, God consulted with the works of every single day. Like a king who had a council of advisers and did nothing without their knowledge. According to Rabbi Ami, God consulted his heart. Like a king who invited a builder to

> construct a palace; if, when he saw the palace, he did not like it, whom should he blame? The builder of course. In the same way God blamed his own heart.

These are all personal opinions, which arose from the desire to make sense of what had been handed down as tradition. The riddles in the ancient texts have still not been solved to this day. The *midrashim* go through the holy books line by line, discussing and interpreting every phrase. These devout scholars dedicated themselves to these texts – they *had* to make sense. To do this, they searched and extrapolated, compared and suppressed. Here is one more example to show you what I mean, this time from the *Midrash Shemot Rabba*. This consists of 52 chapters and deals with the Book of Exodus.

> And God spake to Moses. According to Rabbi bar-Mamal, God said to him: 'Thou wishest to know my name. I am named according to my deeds; sometimes am I called God the Almighty, sometimes Zebaoth, sometimes Elhim; when I wage war against the blasphemers I am called Zebaoth, when I punish men for their misdeeds I am called God the Almighty, and when I show mercy unto the world I am called Jehova, for this name is the very meaning of mercifulness.

30 Beer, B, *Leben Abrahams, nach Auffassung der jüdischen Sage*, Leipzig, 1859

31 Riessler, P, *Altjüdisches Schrifttum ausserhalb der Bibel. Die Apokalypse des Abraham*, Augsburg, 1928

32 Al-Makrizi, Taki, *Das Pyramidenkapitel in al-Makrizis 'Hitat'*, translated by E Graefe, Leipzig, 1911

33 Bonwetsch, N G, *Die Bücher der Geheimnisse Henochs. Das sogenannte slawische Henochbuch*, Leipzig, 1922

34 Kautzsch, E, *Die Apokryphen und Pseudepigraphen des alten Testaments, Vol. 2: Das Buch Henoch*, Tübingen, 1900

35 Riessler, P, *Altjüdisches Schrifttum ausserhalb der Bibel. Das Henochbuch*, Augsburg, 1928

36 In this connection, see Berdyczewski, M J (Bin Gorion), *Die Sagen der Juden von der Urzeit*, Frankfurt am Main, 1914

3

The Return of the Gods

'One is never deceived, one deceives oneself.'
(Johann Wolfgang von Goethe, 1749–1832)

For as long as *Homo sapiens* has been capable of thought, he has feared death. He experiences the cycles of death and rebirth in nature. He sees the stars paling at dawn – and growing bright once more the following night. What lies between death and new life – some mysterious condition of expectation, of looking forward to the next birth? Those who have a conviction that life continues beyond death can find the strength to face death with relative equanimity. Yet fear of death remains; for, as we know from our own experience, hope is a shimmering, elusive thing.

The fear of the individual is also the terror of the masses. Whole nations are afraid of war, of the atom bomb, of the destruction of the environment. Many think with unease and apprehension of the terrible events which are threatened in holy texts: doomsday, or the day of judgement. In the New Testament, St Mark announces (13:24–5):

> But in those days, after that tribulation, the sun shall be darkened and the moon shall not give her light. And the stars of heaven shall fall, and the powers that are in heaven shall be shaken.

His colleague Luke is even more specific; he even lists the warning signs which will precede the Day of Judgement (21:10–26).

> Nation shall rise against nation, and kingdom against kingdom: And great earthquakes shall be in divers places, and famines, and pestilences; and fearful sights and great signs shall there be from heaven ... And there shall be signs in the sun, and in the moon, and in the stars; and upon earth distress of nations, with perplexity; the sea and the waves roaring; men's hearts failing them for fear, and for looking after those things which are coming on the earth: for the powers of heaven shall be shaken.

The Koran also describes these tumultuous events in no less dramatic terms (Sura 82).

> When the heavens show their cracks, the stars are scattered, the oceans mingle with one another; also when the graves turn over and are emptied, then every soul will know what it has done and what it has failed to do.

The day of judgement is even evoked in Gregorian chant, in those simple and yet so wonderful, deeply moving songs which are still sung in Catholic monasteries. The Dies Irae (literally 'day of wrath') is sung during the liturgy for the dead.

At this same time of tumultuous destruction, it is said that the 'judge' of the day of judgement will also appear. In Mark (13:26) we hear:

> And then shall they see the Son of Man coming in the clouds with great power and glory. And then shall he send his angels, and shall gather together his elect from the four winds, from the uttermost part of the earth to the uttermost part of heaven.

Luke (21:28) adds another sentence: 'And when these things begin to come to pass, then look up, and lift up your heads; for your redemption draweth nigh.'

The Apocalypse

Of course, it is only the true and faithful who are going to be saved, the devout, blindly believing adherents of the holy scriptures. But if you ask me *which* holy scriptures, I could not tell you; for every religion in this earthly madhouse believes that only its own holy scriptures reveal the truth. A heavenly judge is prophesied to appear 'on the clouds' to measure the deeds and misdeeds of humanity with an ultimate yardstick. And before the lucky chosen ones are carried off to heaven, the rest of humanity will be whipped, beaten, drawn and quartered.

St John gives us the most riveting description of this in his so-called Revelation, the last text contained in the New Testament. We read there that seven seals will be broken open, and that with every seal new plagues will come to afflict mankind. Trumpets will sound, and with every blast terrible events will occur, in which a third of the ocean is turned to blood, a third of all creatures die, and a third of all ships sink.

But still worse is to come when the third trumpet sounds (8:10–11):

> And there fell a great star from heaven, burning as it were a lamp, and it fell upon the third part of the rivers, and upon the fountains of waters; and the name of the star is called Wormwood: and the third part of the waters became wormwood; and many men died of the waters, because they were made bitter.

Finally the sun and moon are shrouded in darkness, and people are plagued by all imaginable creatures – locusts, scorpions etc – without the relief of being allowed to die. There is no end to the terror: horses with the heads of lions appear on the scene, from whose mouths spew fire, smoke and sulphur.

I have no idea whose brain these nightmares sprang from, or what sort of 'visions' St John suffered. What I do know is that various elements of this apocalypse can be found both in the very ancient texts of Enoch and in the much more recent prophet Daniel (7:1–27).

In contrast to the catastrophes in world history which have occurred up to now, which have been confined to relatively small geographical areas, the apocalypse of St John prophesies world-wide destruction from which no one will be safe, and a final reckoning and judgement.

So where have these ideas come from, these images of a terrible reckoning, with a subsequent redemption for the elect?

And more particularly, what kind of 'all-merciful' God is it who tortures and kills the unbelievers and then lets them roast in eternal hell-fire?

Human imagination does not come up with only beautiful visions; it is equally capable of evoking ghastly scenes. Angry people wish their enemies in hell, and then proceed to imagine hell in its most lurid form. It is also clear that people seek comfort for their earthly sufferings by hoping for a more beautiful world in which things will be better for them. By extension they may also wish that others – the bad, the unjust, the rich, the atheists etc – will get their come-uppance and have their turn of suffering, while they themselves sip the nectar of the gods and bask in the glories of paradise.

O the world is so unfair
for I fare ill, while well you fare.
The world would be much less perverse
if I felt better and you felt worse.

The worse things are in the world, the more people long for a future golden age in which justice and equality reign. Since 'nothing can come of nothing' – not even a golden age – a king of some kind is needed, a ruler, a risen one, a redeemer, a prophet, someone in other words who has the power to clean up this pigsty and sort us all out. This psychologically understandable desire is responsible for all the resurrections, messiahs and prophets that we have been graced with through the centuries. Let me describe a few astonishing examples.

Prophets of Our Times

On 5 January 1945, the 67-year-old seer Edgar Cayce died in Virginia Beach, USA. In a state of trance, the 'sleeping prophet', as he was known, had been able to heal countless people despite never having read a single medical book in his life. In roughly 2,500 'readings', he gave extraordinary pieces of information about the past and the future, as well as about his repeated reincarnations from ancient Egyptian times up to the modern day. Many books have been written about him and his adherents number several million.[1]

In November 1926, in Puttaparthi in the Indian state of Andhra Pradesh, was born a boy by the name of Satyanarayana Raju. His first name roughly translates as 'divine man'. When he was 14, Satyanarayana Raju was bitten by a scorpion; and on waking from a coma that lasted several days, he asserted that he was the reincarnation of Sai Baba, who had been a great Indian holy man in the previous century. Satyanarayana Raju went public when he was 30 years old, and at the age of 36 founded his own ashram. Today Sai Baba receives people and gives talks at his birthplace, 250 kilometres north-east of Bangalore. His is the largest ashram in India. There is also a university attached to it and an excellent hospital. His adherents are thought to number around 100 million. Countless books have been written about him.[2] Each day he accomplishes materializations and miracle healings of all kinds, in front of his followers and also for television cameras. He attributes to himself omnipotence, omniscience and omnipresence, and proclaims himself to be a reincarnation of Buddha, Krishna, Rama and Christ. The German magazine *Der Spiegel* has reported that he is also not averse to physical sex.[3] He has prophesied his own death for the year 2022, but only so as to be promptly reincarnated in the Indian state of Karnataka.

In Graz, Austria, on 15 March 1840, something strange occurred. The 40-year-old music teacher Jakob Lorber suddenly heard a clear voice commanding him to write. Obediently, though initially rather frightened, he took up his pen and, in the following years, wrote down volume after volume dictated by the voice, which he always sensed to be 'in the region of his heart'. The collected works of Professor Lorber encompass no less than 25 volumes – roughly 10,000 pages in total.[4] He described scientific and astronomical details which were not discovered until later; and gave astonishing commentaries on both the Old and New Testaments. His followers number a few hundred thousand people, who are firmly convinced of the truth of his teachings.

Also in the last century, in Qadian, a village north-east of Lahore in what is now Pakistan, the prophet Hazrat Mirza Chulam Ahmad was born. During his life he proved to be a gentle, loving person, gifted at speaking and writing; he founded the Ahmadiyya movement, an Islamic community which still has many followers. Miraculous powers have been attributed to him; his adherents swear that God Almighty had 'woken him to continue the task of all past prophets'. He was thought to be 'the messiah and mahdi for Christians and Muslims', as well as 'the Krishna to Hindus, the Buddha to Buddhists ... and a redeemer of all humanity'.[5]

These are just four of many prophet-figures who have appeared within the last 150 years; whatever you may think of them, they achieved astonishing things. Besides such *positive* prophets and healers, who never harmed anyone, there are a multitude of *negative* figures: prophets of doom who have been telling us for years that we should all have been dead long ago. From time immemorial the idea of the end of the world has been a continuous theme; the world itself, however, does not adhere to it.

Believers and Unbelievers

I have no problem in dismissing the prophesies of charlatans, even those who cloak themselves in the guise of science. They are always easy to recognize by their attachment to the present and to particular ideologies. I do not even have a problem with prophets such as Jakob Lorber, Hazrat Mirza Chulam Ahmad, Edgar Cayce or Sai Baba, even if the latter declares himself to be God. Their astonishing abilities and, if you like, their universal knowledge, can be explained by a modern, reasonable, mathematically deduced theory which was formulated by the French atomic physicist Jean E Charon. It runs as follows. Matter and spirit are inseparably united with one another. In every atom, or to be more precise, in every electron, is contained the total intelligence of the universe.[6] This explains the knowledge of the prophets, even if they themselves are not aware of where it comes from – a contradiction in itself!

But I do have a problem on a quite different plane: with religions which tell us that on the day of reckoning the unbelievers will be drowned, killed, stabbed, poisoned (by 'bitter water'), shot, squashed by earthquakes or eradicated by other such disasters. But *which* unbelievers? Those who do not believe in the Catholic dogmas? Those who have had the misfortune to be brought up in a Christian denomination? Those who are unlucky enough not to have been raised in Arabic or Asian lands? Those who are unaware of the teachings of the Koran or of Buddhism or Hinduism? Those who belong to the Shinto religion of Japan? Or those who do not adhere to the *Book of Mormon*? It looks like our dear Lord God has made quite a muddle of things one way or another!

Almost all religions await a redeemer of some kind, a saviour, a reincarnating messiah. For Christianity, Jesus Christ is this figure, the saviour who redeemed us, 2,000 years ago, from the ominous weight of original sin; yet who, nevertheless, is supposed to return 'throned in clouds' to judge us. Why is it, though, that Jesus became the Messiah for Christians while his own people, the Jews, did not recognize him as such? This is all so confusing and – hardly surprising – accompanied by thousands of long-winded commentaries, that I must just concentrate on the essentials.

Was Jesus the Messiah?

It seems very dubious to elevate Jesus into a Christian or even a Jewish saviour – not only because, contrary to the prophesies, no enduring peace came about after him, but also because the reign of the House of David, which was supposed to last for all eternity, died out thousands of years ago! The 'prophetic' Book of Isaiah is translated sometimes into the present tense – 'Unto us a child *is* born' – sometimes into the future – 'Of the increase of his government and peace there *shall* be no end.' The awaited child could, logically, not yet have been born in Isaiah's time. It is therefore helpful to know that the Hebrew script in which the prophetic texts are written is a purely consonantal form, in which there is no grammatical future tense.[7] To make reading easier, the vowels were indicated by small dots between the consonants. In the original text there existed the imperfect (continuous past) and the perfect (completed past). There was no future form at all. Therefore translators can interpret as they like, which is how the sequential past becomes – abracadabra – a future possibility!

The scholars, of course, disagree about which passages of Isaiah are genuine. Whenever one expert asserts that the original Book of Isaiah has been subject to wholesale restructuring, addition and deletion, another will declare the opposite. These are theological disputes to which I have become accustomed through the years. Nobody knows the truth, yet no other messianic prophesies have been endowed with such universal significance as Isaiah 9:6 and Daniel 7:27.

Whoever wishes, at all costs, to derive a messianic Jesus figure from these vague indications and formulations inevitably comes a cropper when confronted by the historical facts. The life of Jesus was succeeded neither by the appearance of some unique power, nor by a kingdom that lasted for ever. Christian theologians know this of course, which is why they invented a hypothetical 'eternal kingdom' that is supposed to *follow* the day of judgement. What has not appeared up to now is supposed to appear in the future – anything to keep one's hopes up!

Whoever beats a path through the desert of theological argumentation will come to recognize in the old texts a hope directed towards the future, a prophesy of some momentous event that is to take place at some time or other. The prophets and apocalyptic writers imagined this event in a variety of ways. The patriarchal prophets clearly envisage the scenario as taking place upon earth, while the apocalypts imagine it somewhere above the earth. The theologian Dr Werner Küppers makes the following telling remark:

> The light cast by this hope shines upon a dark background; and in its focal point appears the shifting form of a mysterious figure: a human-like Son of Man, the chosen one of righteousness, the star of peace, the new priest, the man, the Messiah. How are we to understand such a combination – a figure of purely coincidental stature who is more than just man yet also neither angel nor God?[8]

Jewish theology holds fast to the Messiah as a 'man of human descent';[9] he is often depicted not as an individual personality, but as the entire people of Israel itself. Christian theology sees him differently: as a messianic figure equated with the 'son of God'. But both theological versions leave various questions unanswered. Where did the idea of a messiah originate? How old is it? There is not much point in citing prophets like Isaiah, Daniel or Ezekiel when one knows that their texts have been tampered with and rewritten. Neither, for the same reason, can one rely on them for any kind of accurate dating: the idea of a messiah is clearly much older than the prophets. What they have recorded are only the traces in folk memory of an expectation that has existed since the expulsion from paradise. The prophets and their later editors drew upon the traditional wisdom which encompassed the hopes and expectations of a whole people. This hope was already an integral part, perhaps even the central preoccupation, of a race of human beings, before any words were written down. Expectations of being saved and delivered are 'very ancient, long pre-dating the prophets'.[10]

'The Israelis have bequeathed three gifts to the world,' writes the theologian Leo Landmann,'monotheism, moral edicts and the true prophets. To this must be added a fourth: belief in the Messiah.'[11] This is easy to disprove; many old cultures and peoples had messianic expectations.

In 1919 the theologian H W Schomerns wrote:

> The certainty of Christianity's superiority, indeed its absolute validity, over all other religions, strengthens and edifies the Christian populace.[12]

I think such assertions should be tempered with a knowledge of other religions. One should first read and feel one's way into them; and whoever, after such study, still credits Christianity with absolute superiority, is shutting his eyes and relying on blind faith. Faith is a matter for the individual. Personally I respect the beliefs of every single person. But I think it is wrong to underestimate other religions: they have retained their intensity and fascination for thousands of years – in many cases for longer than Christianity. All religions, whether pre- or post-Christian, contain the idea of redemption. All without exception longingly await the heavenly signs and the promised return of their messiah. The greatest and surely the most dynamic of post-Christian religions is Islam. In the holy book of the Muslims, the Koran, Jesus is honoured as a prophet, but not revered as Messiah or son of God.

The Messiah of Islam

Christianity is alone in believing Jesus to be the Messiah and Redeemer. None of the other great world religions adheres to this belief, neither Judaism nor Islam, let alone the religions of Asia.

Now all these world religions had, and still have, their own excellent researchers, thinkers and exegetes. All of them had, and still have, first-rate colleges and places of learning, staffed with armies of multilingual experts. But to me, as a theological layman, it seems astonishing that on the basis of *the same material*, all these super-intelligent egg-heads arrive at wholly different versions of the truth. Judaism, Islam and Christianity all base their exegeses on *the same* ancient prophets. So how can it be said that exegesis is an exact science? If this were so, surely one might expect them to arrive at similar results. Since this is clearly not the case, I say that no one knows the truth any more. These researchers just serve their own cause, whether they believe in it or not.

Islam also contains the idea of the day of judgement and the final reckoning. Similar to the Revelation of St John, the Koran tells us (Sura 21, Verse 105):

> Upon the day when We roll up the heavens, as documents are rolled up. As We began the first Creation, so we shall renew it …

Or, similar to the trumpets in Revelation, another verse from the Koran (Sura 20, Verse 103) says: 'Upon the day when the trumpet shall sound. Upon that day We shall gather together the guilty, blue-eyed ones.' Sura 17, Verse 59, even remarks that no town will remain standing on the day of punishment and resurrection.

And when is this supposed to happen? That is the secret of Allah (Sura 21, Verse 41):

> No, it will come upon them unexpected, so that it casts them into confusion; nor will they be able to defend themselves from it, nor will they be granted any delay.

The Islamic messiah is called 'the Mahdi'. Both the prophet Mohammed and the various imams who followed him proclaimed the return of the Mahdi. The imams – the great teachers of Islam – always held it to be wrong to speculate about the date of the Mahdi's return, for this was a secret known only by Allah. Just as in Judaism and Christianity, the literature about the second coming of the Mahdi fills whole libraries. There is nothing on this subject that has not already been thought and written down by someone. A foreigner once enquired of the fifth imam, al-Baquir, what signs would be witnessed before the Mahdi's return. He replied:

> It will happen when women behave like men, and men like women; and when women sit with spread legs upon saddled horses. It will happen when false prophesies are held to be true, and true prophesies are rejected; when men spill the blood of other men for little purpose, when they do indecent acts and scatter and waste the money of the poor.[13]

According to these criteria, the Mahdi is long overdue. Not to mention that, before the Mahdi comes, '60 false men will appear, who make themselves out to be prophets'. By my reckoning there must have been a good deal more than 60,000 false prophets so far.

The same theological chaos exists in respect of the return of the Mahdi as we find about the Messiah in Judaism and Christianity. The great world religions all expect a messiah, but no one knows when he will arrive. This messiah figure is usually seen in connection with the stars, the firmament and the ultimate reckoning of human deeds. He is supposed to be accompanied by hosts of angels, to possess immense power, and to be throned upon the clouds. Do these beliefs derive from a core of folk memory? Do they recall a primeval promise of 'We will return'?

To make these vague suppositions more concrete and precise, we need to turn to older and other traditions than those of the Koran or the Christian apocalypse.

The little word *Avesta* comes from Middle Persian and means basic text or instruction. The *Avesta* contains the complete religious texts of the Parsees, or the modern adherents of Zarathustra. Zarathustra is supposed to have been virginally conceived. Tradition has it that a mountain arrayed in pure light sank down from the skies. From the mountain emerged a young man, who implanted the embryo of Zarathustra into the womb of his mother. Because their religion was older than Islam, the Parsees refused to accept the Koran as their holy book. They emigrated to Iran and India. Although their language, Gujarati, is a modern Indian language, they continue to conduct their worship in the temple language of the *Avesta*, comparable to the Catholic tradition of holding services in Latin.

The Parsees are in a similar dilemma to the adherents of other religions: only about a quarter of the original texts of the *Avesta* are still extant. Portions of this ancient Persian religion were retained in cuneiform script, which King Darius the Great (558–486 BC), his son Xerxes (about 519–465 BC) and his grandson Artaxerxes (about 424 BC) ordered to be made. The highest god of this religion is called Ahura Mazda, and he was the creator of heaven and earth.

Praised Be the Stars!

In the Parsee texts, the fixed stars are ordered in various star groupings, each of which is led by particular 'commanders'. The heavenly hosts are a decidedly military bunch; there are 'soldiers' of the constellations, and also battles conducted throughout the universe. The different stars are praised in the highest terms (*Afrigan Rapithwin*, Verse 13):

> The star Tistrya, the shining, majestic one we praise.
> The star Catavaeca, which rules the waters, we praise.
> All stars which contain water-seeds, we praise.
> All stars which contain the seeds of trees, we praise.
> Those stars which are called *Haptoiringa*, the healing ones, which oppose the Yatus, we praise ... [14]

These tributes seem to be more than just ornamentations of pure fantasy, for the Parsees had, from the beginning, some degree of astronomical knowledge. The planets, for example, were known to them as 'simple bodies of round form'. From earliest times the Parsees' temples had honoured the various gods and their places of origin in the universe, in ways which almost prefigured the revolution in astronomical thinking brought about by Galileo Galilei in 1610. In every temple could be found a round model of the planet to which it was dedicated. There were specific kinds of clothing and customs in each temple, depending on the planet it honoured. In the temple of Jupiter, one had to appear in the dress of a judge or scholar; in the temple of Mars, on the other hand, the Parsees wore red, martial dress and had to converse in 'proud tones'! In the temple of Venus one laughed and joked, in the temple of Mercury one was meant to speak like an orator or philosopher. In the temple of the moon, the Parsee priests behaved like childish wrestlers, jumping and rolling about the place. In the sun temple one had to wear brocade and behave 'as befits the kings of Iran'.

The *quadriga solis*, the four-horse chariot with winged steeds, originates in Iranian folklore;[15] in the Parsee version, the gods of particular planets take it in turns to drive the sun-chariot. And in the texts of the *Avesta*, the heavenly chariot and its drivers are praised in the following terms (*Yasna*, Chapter 57, Verse 27):

Four steeds,
white, bright, shining,
shrewd, wise, shadowless,
ride through the heavenly regions ...
faster than the clouds,
faster than the birds,
faster than arrows,
they overtake all
who follow behind them ...

In these texts the universe abounds in such flying machines. The Parsees also, it almost goes without saying, looked forward to the reappearance of their gods. They believed that 'light-beings'[16] would descend from the heavens and save suffering humanity. Zarathustra himself questioned his god Ahura Mazda about the end of the world, and was told there will be a final battle of the good against the corrupt. From the heavens will descend many 'all-conquerors'. These will be immortal and possess knowledge of all things. Before they appear in the skies, the sun will be shrouded in darkness, there will be earthquakes and mighty storms and winds, and a star will fall from heaven. After a terrible battle, in which armies confront each other in hosts, a new, golden age will dawn. Humanity will then become so knowledgeable in the arts of healing that 'they can cure one another, even when close to death'.

This version of 'redemption' does not seem to be all that different from the one we find in other religions, except for the fact that it is these 'all-conquerors', the gods from the starry worlds, who appear as final, long-awaited saviours.

The Golden Age

In Hinduism everything is more complicated because of the manifold deities. At the beginning of the four world epochs was an Age of the Gods, the *Krtayuga* or *Devayuga*. In all respects this period was perfect, for there existed neither illness nor envy, neither dispute nor ill-will, neither fear nor pain. In those days, according to Hindu teachings, the aim of all people was fixed solely on the highest Brahma, and even the members of the four castes lived in harmony with one another. Life and human beings themselves were simply perfect. People devoted themselves to an ascetic life and the study of the scriptures. Material desire was unknown. People loved truth and knowledge. There was no injustice, for no one felt any earthly longing. The *Bhagavata-Purana*, one of the many works of Hindu religion, describes the people of that golden age as content, friendly, patient, gentle and merciful. They were happy because they bore peace within their hearts and were not at odds with anything.

It was therefore a world which we can hardly imagine. Nowadays, of course, we are torn hither and thither by desires and wishes. The idea of an age of absolute happiness uninformed by desires is quite foreign to us. Yet this golden age of Hinduism is, so to speak, only a wish projected into the far distant future. As it was in the 'dream age', so it will be once more in the future. A time of beauty, strength, youth and harmony will return.

Hinduism does not have a 'founding' couple like Adam and Eve; Brahma created 8,000 people at one go – 1,000 couples of each caste – who were like the divine beings. These couples loved one another and were united with one another, yet did not produce any children. Only at the end of their lives did these pairs bear two children each; not through sex, but through the power of thought alone. In this way the earth was populated with spiritual beings.

This happy state of affairs lasted until negative spirits, as well as gods of all kinds, introduced chaos and confusion amongst human beings. The gods were seen as hugely powerful and immortal beings, yet in other respects similar to humans, and endowed with individual personalities. Highest of these deities was the 'Prince of the Universe, who ruled over all'.[17] The Hindu gods are so many, various and interrelated that I cannot describe them here in greater detail. Suffice it to say that the gods had mastered both air and space travel by means of flying machines of all sorts and descriptions. All these flying objects were of a real, material nature – they were not spiritual, nor did they arise from fantasy or imagination.

Flying apparatus with alarming weapons systems are described in Indian religious texts in great detail, particularly in the Vedas, which are thought to be the most ancient sources of language and religion. The word *veda* means 'holy knowledge'. One of these texts, the *Rigveda,* is a collection of 1,028 hymns to the gods. It states in no uncertain terms that these flying machines came from the cosmos to earth, and that the gods came in person to impart knowledge to human beings. Similar to the Jewish legends, the Hindu texts describe battles among the gods; not, however, in some undefined heaven of spiritual glory, but 'in the firmament', 'above the earth'.

Star Wars

In the 'Vanaparvan', which belongs to the ancient Indian *Mahabharata* (Chapters 168–73), the dwellings of the gods are described as space settlements, which orbited high above the earth. The same sort of thing can be found in Chapter 3, Verses 6–10, of the *Sabhaparva.* These gigantic space stations had names such as Vaihayasu, Gaganacara and Khecara. They were so enormous that the shuttle-ships – the *vimanas* – could fly right into them through mighty gates.

We are not talking about obscure fragments which no one can examine, but ancient Indian traditional texts which are to be found in any large library. In the 'Drona Parva' section of the *Mahabharata,* page 690, Verse 62, we can read how three great and beautifully built cities revolve around the earth. From these, discord spreads to the people on earth; and also amongst the gods themselves, in a war of galactic proportions (Verse 77).

> Siva, who rode upon this most excellent chariot, that was composed of all the forces of heaven, prepared himself for the destruction of the three [heavenly] towns. And Sthanu, this leader of the destroyers, this thwarter of the Asuras, this fine fighter of immeasurable bravery, drew up his forces in an excellent battle-position ... When the three towns next crossed each others' paths in the firmament, the god Mahadeva shot them through with a terrible stream of light from the threefold mouth of his weapon. The Danavas were unable to look up into the path of this streaming light, which was ensouled with yuga-fire and contained the power of Vishnu and Soma. While

> the three settlements began to burn, Parvati hurried there to see the show.[18]

The gods of Hinduism battled with each other 'in the firmament' like Ishmael (or Lucifer) in Jewish tradition:

> Ishmael was the greatest prince of angels in heaven ... And Ishmael went and united with all the highest armies of heaven against his Lord; he gathered his armies about him and descended with them, and began to seek companions upon the earth.

And what do we read in Enoch? He described the mutiny among the angels, and even listed their names.

This core of tradition – the battle in heaven, the struggle between the gods – is the decisive thing, and is made a farce of by the naive concept of heaven to which the various religions subscribe.

In Hinduism, human beings attain absolute serenity through their own powers, through continual cycles of rebirth during which they improve and cleanse their *karma*. Yet they are helped in this by the gods, and ultimately by the universal god Brahma. But the Hindus are also familiar with the idea of the gods' return. Vishnu will one day be reborn as Krishna and save the earth from the mess it has got itself into. Where the idea of *karma* or reincarnation fits into all this is a mystery for westerners. How did the Hindus ever come to believe in a continuous cycle of rebirth, in which they haul their good and bad deeds from one life to the next?

The extraordinarily complex teachings of *karma* are described in the Jain religion in very precise detail. Jainism is, alongside Buddhism and Hinduism, one of the three major religions of India. Centuries before the arrival of Buddhism, Jainism arose in northern India, then gradually spread throughout the whole subcontinent. Its adherents say that it was originally founded in very ancient times – millennia ago. They believe its teachings are eternal and imperishable, even though they may be forgotten for long periods at a time. The Jain religion is encompassed in a whole series of pre-Buddhist texts, which are – there is no other way of putting it – quite extraordinary.

Ancient Science

The theological-cum-scientific literature of Jainism contains stories of holy men, songs about the primeval creators, as well as precepts of all kinds. These works are – in a similar way to the Bible – collected together under the umbrella of a single title: *Shvetambaras*. They are divided into 45 main sections, each with quite unpronounceable names.

The 'Vyahyaprajnaptyanga' presents the whole teaching of Jainism in dialogues and legends. The 'Anuttaraupapatika-dashanga' tells the stories of the primeval holy ones, who rose into the highest heavenly worlds.

In the 'Purvagata' section are scientific books and descriptions. Within this, the 'Utpada-Purva' deals with the formation and dissolution of all the different substances (chemistry). The 'Viryapravada-Purva' describes the forces active in the substance of gods and great men. The 'Pranavada-Purva' examines the art of healing. The 'Lokabindusara-Purva' deals with mathematics and redemption.

As if all that was not enough, there are also the 12 'Upangas', which describe all aspects of the sun, moon, and other planetary bodies, as well as the life-forms which inhabit them. In addition, the 'Aupapatika' tells us how divine existence can be attained. We are also provided with a list of divine kings (*Prakirnas*, Book 7).

Apart from these writings, there are also supposed to have been books which existed in the deep, primeval mists of time, but which have been lost. But the Jains believe that such writings were passed on orally, from priest to priest, down through the generations. They are not perturbed by their loss, since reincarnations of the ancient prophets continually reappear, who reveal their content anew – to the extent that the times and people are ready to receive such teachings. The content of the lost texts has only survived in fragments, but even these deal with the most astonishing things:

- how one can travel to far lands by magic means
- how one can perform miracles

- how one can transform plants and metals
- how one can fly through the air

Flying through the air is also described in Sanskrit literature. My book, *Der Götter-Schock,* deals with this in detail.[19]

According to the Jain teaching, the epoch in which we live is only one of many. Before our time there were other cosmic periods; and shortly – round about the year 2000 – a new epoch is supposed to begin. Such new epochs are always heralded by 24 prophets, the *tirthamkaras*. The prophets of our time are just being born, or are perhaps already adults. The religious leaders of Jainism believe that they even know their names and other details of their lives.

Impossible Dates

The first of these *tirthamkaras* was Rishabha. He dwelt for a staggering 8,400,000 years upon the earth. Rishabha was of giant proportions. All the patriarchs who succeeded him gradually diminished in stature and longevity; however, the 21st of them – whose name was Arishtanemi – still lived for 1,000 years and was ten bow-lengths tall. Only the two last, Parshva and Mahavira, lived to what we would consider a 'reasonable' age. Parshva lived to 100 and was only 9 feet tall, while Mahavira, the 24th of the *tirthamkaras,* only made it to a mere 72 and was just 7 feet tall.

The Jains place the appearance of their *tirthamkaras* in such ancient times that it makes one dizzy to think of it. The two last were supposed to have died in 750 and 500 BC respectively, while the successor to Rishabha (the first patriarch) graced the earth for about 84,000 years.

These numbers, which are just set down in front of us, should actually make our myth investigators, and our theologians as well, sit up and take notice. Why? Because we have here, well packaged in religious concepts, a kernel of folk memory that gleams through many holy and not-so-holy books. Let me, very briefly – in telegram style – refresh your memory.

In the ancient Babylonian list of kings (WB 444), ten kings are counted from the creation of the earth up to the flood. These ruled, give or take a year or two, for 456,000 years. After the flood 'the kingdom descended once more from the heavens'[20] and the 23 kings which followed ruled altogether for 24,500 years, 3 months and 3 1/2 days.

To the biblical patriarchs are attributed just as unbelievable ages. Adam is supposed to have lived for more than 900 years; Enoch was 365 years old when he rose up into the clouds; and his son Methuselah carried on for 969 years.

It was no different in ancient Egypt. The priest Manetho recorded that the first divine ruler in Egypt was Hephaistos, who also brought the gift of fire. Then followed Chronos, Osiris, Tiphon, Horus and the son of Isis.

> After the gods, the race of god-descendants ruled for 1,255 years. And then other kings ruled for 1,817 years. After this, 30 more kings ruled for 1,790 years. Then still 10 others for 350 years. The kingdom of the spirits of the dead and the god-descendants encompassed 5,813 years.[21]

Such impossible dates are confirmed by the historian Diodor of Sicily, who, 2,000 years ago, wrote a whole library of works consisting of 40 volumes:

> From Osiris and Isis until the rule of Alexander, who founded the city in Egypt which is named after him, more than 10,000 years are said to have passed; some say, though, that the period is actually only a little less than 23,000 years ...[22]

And as my last witness testifying to such impossible dates, let me mention the Greek Hesiod. In his *Myth of the Five Races of Mankind*,[23] he wrote – about 700 BC – that originally the immortal gods, Chronos and his companions, created human beings: 'Those heroes of excellent descent, called demi-gods, who in the time before ours dwellt upon the endless earth ...'

Let us now return to the Jains – who, it has turned out in the meantime, are by no means alone in recording dates of staggering proportions. Many Jain accounts are – seen from the standpoint of modern science – quite revolutionary. Their concept of time, of *kala*, seems like something formulated by Albert Einstein.

Their smallest unit of time is the *samaya*. This is the time it takes for the slowest atom to move the distance of its own length. Innumerable *samayas* form one *avalika*, and – measurable at last – 1,677,216 of these *avalikas* give us one *muhurta*, which equates with 48 of our minutes. Thirty *muhurtas* give us one *ahoratra*, which is exactly one day and night. Do you get it yet? Multiply 48 minutes (one *muhurta*) by 30, and you arrive at 1,440 of our minutes, which is exactly the same as the number of minutes in 24 hours. But the time-reckoning of the Jains is thousands of years old, and was originally passed to humans by heavenly beings.

Fifteen *ahoratras* give – conforming with our time measurement – one *paksha*, which is half a month; two *pakshas* are therefore of course one *masa*, or one month. Two months make one season, three seasons make one *ayana*, or term. Two *ayanas* give us one year, and 8,400,000 years make one *purvanga*. But it continues: 8,400,000 of these *purvangas* make one *purva* (=16,800,000 years). The Jains' counting carries on to 77-figure numbers. Beyond this, their time-values are given in terms of specific concepts, similar to our light years, for a distance of 9,500,000,000,000 kilometres.

One might well be tempted to call this crazy idiosyncrasy, if it were not for the fact that the Mayas of Central America operate with similarly staggering numbers, and also relate them to time and the universe in the same way as the Jains in far-distant Asia.

The Jains also took from their heavenly teachers definitions of space which are astonishing, and which in the end – or at last? – render comprehensible its connection with the mysterious idea of *karma*. I can only give a brief resumé here of this extremely complex and involved doctrine, an understanding of which I gleaned from a book by the theologian Helmuth von Glasenapp.[24]

In the scientific writings of the Jains, the atom occupies a point in space. This atom can connect with others to form a *skandha*, which then encompasses several or an immeasurable number of spatial points. Our own science teaches the same: two atoms can form a chain of the smallest proportions; but there are also molecule chains consisting of many millions of atoms. These atom chains give rise to substance and materials of various densities. The Jain teaching distinguishes six chief forms of such chains or connections:

- fine-fine: things that are invisible
- fine: things that are still invisible
- fine-coarse: things that are invisible but perceptible by smell and hearing
- coarse-fine: things one can see but not feel, such as shadow or darkness
- coarse: things which reunite by themselves, such as water or oil
- coarse-coarse: things which do not reunite without external help (stone, metal)

In Jainism, even a shadow or a reflection is considered material, because it is produced by a *thing*. Even sound is not categorized as 'fine-fine', but as fine materiality, which arises as a result of 'aggregates of atoms rubbing against one another'.

In this teaching, the 'fine-fine' substance is able to permeate everything and therefore have an altering influence on other substances. The substance which penetrates a soul expresses itself as *karma* – which brings us back to rebirth. Still with me?

Karma *Remains Eternal*

It is common knowledge nowadays that every kind of matter – whether a table or a bit of bone – can be reduced to the atomic level. The atom itself consists of sub-atomic particles, one of which is the electron, which oscillates in an unimaginable rhythm of 10 to the power of 23 per second. The matter of this electron would be characterized by the Jains as 'fine-fine'; it can no longer be grasped and, in addition, is immortal. Atoms can enter all possible chains and combinations, but the electron always accompanies them. It acts like the 'spirit within matter',[25] similar to a magnetic field or radio wave, which permeates particular substances. Now the thoughts of every life-form influence its deeds. 'The substance of the world is the substance of the spirit,' wrote the English astronomer and physicist Arthur Eddington (1882–1944). And the Nobel prizewinner Max Planck formulated it in these words:

> There is no matter as such! All matter arises and is sustained only by virtue of a force which makes atomic particles oscillate.

Our existence is the consequence of a previous act. We would not exist without a prior life which gave rise to us. (And this will not change if, in the future, we learn how to create life artificially.) In other words, every existence is a link in the long chain of previous or future existences. Since our thoughts direct our actions, these actions in turn leave their traces upon our mind or spirit. One could, for example, describe a magnetic field as mind, but it is one which has an influence on matter. The Jains view what we call 'soul' as the 'fine-fine' materiality of the physical body. This materiality is as untouched by the body as the electron is by the atom nucleus. An electron does belong to the atom, but the two never come into actual contact. The atom can alter its position, join with others to form gigantic molecule chains, and will always be accompanied by electrons – but, strange to say, not the *same* electrons, for the electron 'leaps' from one atom to another, for example when heat is applied. And in the very billionth of a second in which an electron leaps to a new atom, the place vacated by it is filled by another electron. So we have an eternal, immortal 'fine-fine' activity, an oscillation beyond the material atom.

The Jains see *karma* in the same way. It does not matter what happens to the body – whether it is burned or eaten by worms – for *karma* remains immortal. This *karma* contains all the information about the life-form to which it belongs. During life we think and feel; this thinking and feeling is transposed onto the 'fine-fine' substance of *karma*, similar to an engraving. When this *karma* is formed into a new body, it already contains all the information from its previous existence, and continues to do so for all eternity. But since the ultimate aim of life is to attain a condition of absolute serenity – by becoming one with Brahma – *karma* will lead us towards this goal through a series of countless reincarnations.

This way of thinking is not so far removed from modern philosophy and the discoveries of modern physics. What may astonish us, though, is that such complex theories were taught millennia ago, and by teachers who appeared out of the depths of the universe. The last epoch of the Jains (which was followed by our own times) began around 600 BC with the last of the 24 *tirthamkaras*. This *tirthamkara* was called Mahavira; and who was he? A king's son whose embryo was implanted in the womb of his mother, the young queen, by heavenly beings.[26] All these heavenly teachers of ancient times are supposed to reappear, born into new bodies. The Jains have many old drawings which depict the 24th *tirthamkara*, the prophet Mahavira. Above the procession in his honour, portrayed in the engraving shown in the plate section, float five heavenly aircraft.

There is a distinct difference between the Jains' expectations of the return of the gods, and those of Christians, Muslims or Jews. The latter believe that a messiah and highest judge will appear, after which the faithful will enjoy heavenly glory while the unfaithful roast in hell. The Jains do not expect a single saviour, but several at once. The prophets or *tirthamkaras* continually return, at every epoch. There is no final end of the world after their appearance – not heavenly joy and ambrosia, nor eternal damnation either, but simply a new act in the drama of the universe. The *tirthamkaras* are less saviours than helpers. They prepare human beings for the next stage and epoch. That is why they are born as human beings (think of the 'son of man' in the prophesies of Enoch); but their substance, their karmic knowledge, derives from the universe. Not earthly but extraterrestrial forces plant the seed or the embryo into the womb. It is also worth remembering that these ideas were current centuries, if not millennia, before the birth of Christ, and that the Jains can therefore hardly have taken the immaculate conception from Christianity – rather the other way around!

It is not surprising if such cosmic teachers as the *tirthamkaras* were well versed in astronomy and astrophysics. It is from such a source that the Jains derived their – to us incomprehensible – astronomical dates. Their teachings show that they were able to measure the dimensions of the universe. Their unit of measurement was the *rajju* – the distance which God flies in six months, when he travels at 2,057,152 *yojanas* a second.

The Jain teachings say that the earth is surrounded by three layers, which are characterized according to their density: dense as water; dense as wind; and dense as a fine wind. Beyond these lie empty space. Our modern science has come to the same conclusion: atmosphere; troposphere, containing nitrogen and oxygen; and stratosphere with the ozone layer. Beyond that is interplanetary space.

People nowadays have increasingly come to hold the view that other life-forms apart from us must exist in the universe. The Jains have always believed this; for them, the whole universe is filled with forms of life which are distributed unevenly across the heavens. It is interesting to note that though they recognize the existence of plants and basic life-forms on many different planets, it is only on a few specific planets that beings exist who are endowed with 'voluntary movement'.[27]

The philosophers of the Jain religion describe the different characteristics which the inhabitants of various worlds possess. The heavens of the gods even have a name: Kalpas. There one can, apparently, find wonderful flying palaces – moving structures of which whole towns are often composed. These heavenly towns are ranged one above the other in such a way that from the centre of each 'level' the *vimanas* (divine chariots) can venture forth in all directions. When one epoch ends, and new *tirthamkaras* are due to be born, a bell sounds in the chief palace of 'heaven'. This bell causes bells to ring in all the other 3,199,999 heavenly palaces. Then the gods gather together, partly out of love for the *tirthamkaras*, partly out of curiosity. And then, borne on a flying palace, they visit our solar system; and a new epoch begins upon the earth.

Waiting for the Super-Buddha

In Buddhism, the fundamental idea of redemption appears in a very similar form to Jainism. The Jains, however, were teaching *before* the arrival of Buddha (560–480 BC). *Buddha* means 'the awakened one' or 'the illumined one'. Buddha's ordinary name was Siddartha. He came from a noble family and grew up in the lap of luxury in the palace of his father, in the foothills of the Nepalese Himalayas. At the age of 29, he had had enough of his rarefied existence. He left home, devoted himself for seven years to the art and practice of meditation, and sought a path of knowledge.

But the gods of folklore, legend and mythology had already been around for a long time in Buddha's day. After his illumination he felt himself to be the incarnation of a heavenly being. He began to preach to his disciples the fourfold path, which could lead all people to buddha-hood, to illumination. Buddha was convinced that the future would bring other buddhas. In his farewell speech, the *Mahaparinibbana-Sutta*, he speaks of these buddhas of the future. He prophesied to his disciples that one of them would come at a time when India would be crammed full of people; the towns and villages would then be as densely populated as hen-coops. In the whole of India there would be 84,000 towns; in the town of Ketumati (the modern Benares) would live a king by the name of Sankha, who would rule the whole world, but without force, just through the power of his righteousness. And during the rule of this king, the sublime Metteya (also called Maitreya) would descend to the earth – a phenomenal and wholly unique 'chariot-driver and knower of worlds', a teacher of gods and men: in other words, the perfect Buddha.

Buddha's prophecy of a 'super-Buddha' is similar to the Jain teaching of the return of the *tirthamkaras*. Buddhism also speaks of the different epochs, which are compared with a turning wheel. The only difference is that in Buddhism these epochs are immeasurably long.

The idea of four – or in Jainism, six – epochs also informs Sumerian-Babylonian mythology. In cultures which are far removed from each other, the same numbers are often found. A professor of religious history, Dr Alfred Jeremias, became aware of such parallels 65 years ago. Here is just one example.[28]

According to Babylonian accounts, the ancient kings or rulers of heaven ruled for thousands of years. The dates applied to the gods Anu, Enlil, Ea, Sin and Samas are remarkably close to the periods attributed to the *yugas* or epochs in India:

Anu	= 4,320	Kali-Yuga	= 432,000
Enlil	= 3,600	Kali-Yuga	= 360,000
Ea	= 2,880	Deva-Yuga	= 288,000
Sin	= 2,160	Treta-Yuga	= 216,000
Sama	= 440	Dvapara-Yuga	= 144,000
Adad	= 432	Maha-Yuga	= 4,320,000

There is a reason for the fact that Kali-Yuga appears twice; the Kali-Yuga 'without twilight' is of shorter duration than the Kali-Yuga 'with twilight'. The number of zeros is not important, but the correspondence of the basic digits demonstrates a common original source. The number 4,320,000 of the Maha-Yuga ('great epoch') is identical with that of the third pre-diluvian king En-me-en-lu-an-na, who ruled for 12 *sar* or 43,200 years. And the number 288,000 of the Deva-Yuga corresponds to the period of rule of the sixth king, En-sib-zi-an-na. He survived for eight *sar*, or 28,800 years.

In Greece can be found the oldest literary reference to a world epoch, in the writings of the poet Heraclitus. He quotes a period of 10,800,000 years, which corresponds precisely with the second period of the ancient kings of Sumeria – 30 *sar*, or 108,000 years.

These numbers do not have any direct connection with the return of some saviour or other, but they demonstrate the common ground that different traditions share. The only way to explain these correspondences is to assume that there must, at the dawn of time, have been a single original teaching. This common source must lie very far back in ancient times, for otherwise it would be mentioned in historical records.

Psychological Cover-ups

Psychology does not help me in the slightest in my researches into the idea of the gods' return. I have ascertained that all cultures demonstrated this idea in some form, and that it is always connected with the stars and with saviours who come from beyond the earth; in addition, frequent mention is made of artificial fertilization of an embryo that is brought by the 'gods'. I have no choice but to believe that these ideas have a common origin, and one which is inaccessible to psychology. Of course it is understandable that people might long for a great saviour, king and 'super-Buddha' – when times are bad enough, people wish for all sorts of cloud-cuckoo lands. But this cannot explain the connections and correspondences between all the different traditions. Wishes alone cannot provide such precise accounts in the first person, and all the details of dates and names. Or do people think that Enoch invented the long list of names and functions of the mutineer 'angels'? Or that the idea of measuring the universe with the number 2,057,125 *yijanas* simply floated down into the head of some dreamer under a fig tree? Psychology is of no more help, either, in explaining the identical dates among different cultural traditions, or the widespread idea that artificial fertilizations and embryo-implants took place. Quite another matter is the way that later religions transformed these concepts in order to glorify their saviours with an immaculate conception – that, certainly, is understandable from a psychological point of view.

Even today, Catholic Christians still believe that Jesus was immaculately conceived by Mary. They have to believe it, for it is a dogma (or article of faith) of the Church. Though to be strictly fair one should add that the opposite cannot be scientifically proved either. How can we actually *know* whether Jesus, or, if you like, the Indian prophet Sai Baba, did *not* develop from a cosmic seed? That is, after all, what went on in ancient times: all great gods and god-kings had to have immaculate credentials so as to be thought equal to their predecessors.

Seeds From Heaven

The seed which developed into the Accadian king Hammurabi (1726–1686 BC) was said to have been implanted in his mother by the sun-god. Hammurabi later became the greatest law-giver. From him derive the most ancient recorded rules and regulations for ordering human society: the *Codex Hammurabi*. The stone pillar, over 2 metres high, upon which these laws were engraved was dug up at the beginning of our century in

Figures 1–3 (above and overleaf)
These mixed-breed beings are engraved on the black obelisk of the of the Assyrian king Salamasar II. They must have been alive, for their guards are holding them on a short leash. (British Museum, London)

Figure 4 (left)
The small creature being led away by King Salamasar III is supposed to be 'an ape'. Neither the feet nor the spread fingers belong to an ape.

Figure 5 (below)
These creatures of fables are also obviously kept on a short leash. (Make-up tablet of King Narmer, The Egyptian Museum, Cairo)

Figure 6 (opposite)
This is how NASA imagines future 'space-habitats' in which journeys of light years could be made.

Figure 7 (above)
On this old engraving of the Jain religion, the 24th *tirthamkara* is worshipped in a procession. Above float airships.

Figure 8 (right)
The seed of the prophet *tirthamkara* is placed by artificial insemination into the womb of the young princess.

Figures 9–12 (opposite and overleaf)
In the middle of Java stands the largest Buddhist temple, the Borobudur. The wall reliefs show scenes from the life of Buddha. Everywhere stand *stupas* which are here seen as a means of transport to the divine worlds. That is why a young buddha sits in every *stupa*.

Figures 13–14 (opposite) The oldest known representations of the gods are to be found on Sumerian scroll-seals. Portrayed are scenes in which the gods impart knowledge to the kings, and scenes of unknown 'mixed beings'.

Figure 15 (above)
Detail from a Sumerian scroll-seal showing a divine chariot in the firmament.

Figure 16 (opposite)
Representation of a weapon of the gods, the 'thunder-hammer', in the Kva Bahal Monastery in Nepal.

CHOSES EN
革の

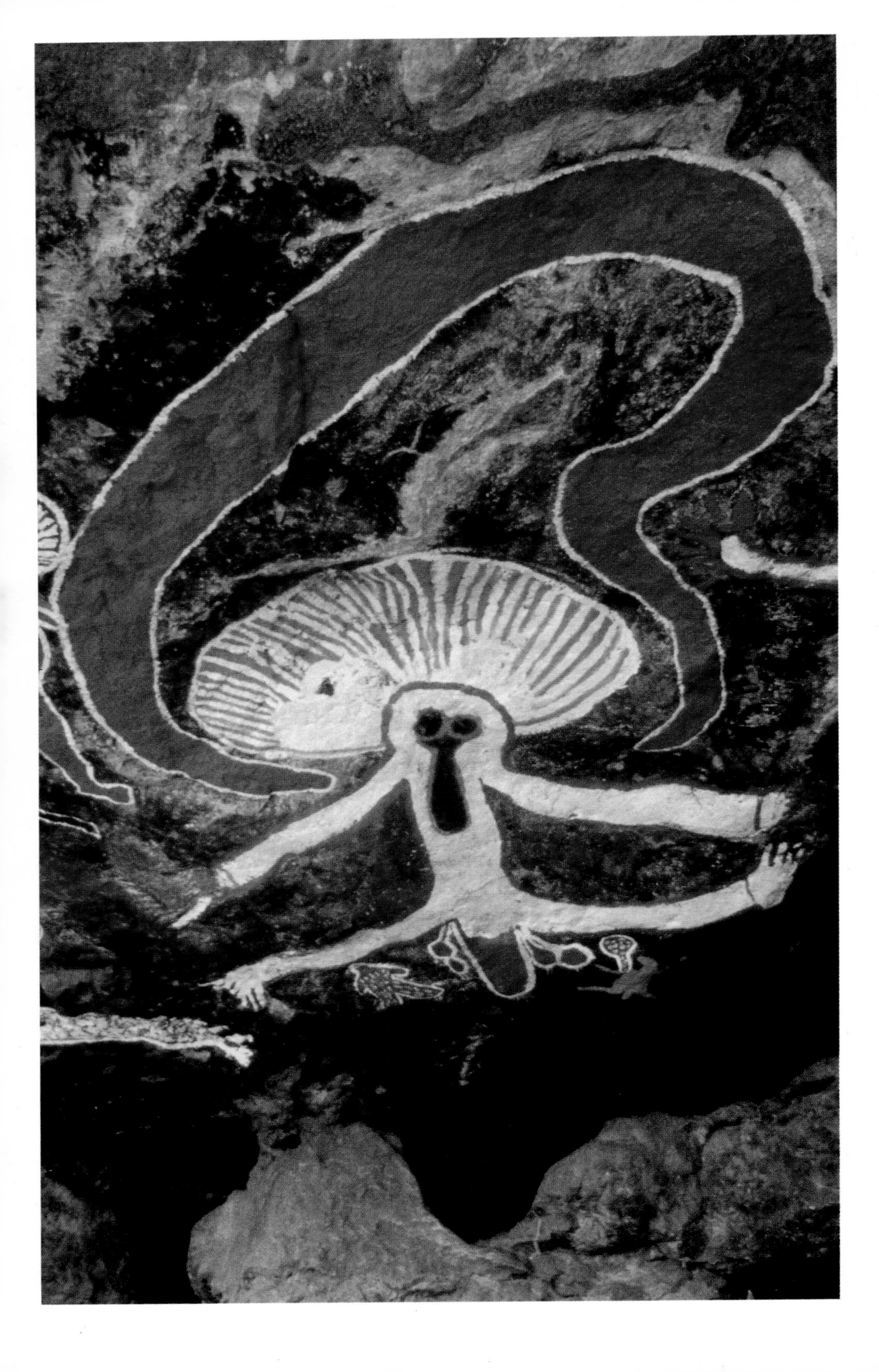

Figures 17–18 (left and above)
Cave drawings of the Australian Aborigines depicting the divine goddess Vandina surrounded by streaming rays.

Figure 19 (overleaf)
The Pyramids of Giza.

Figures 20–21 (above and left)
The robot Upuaut constructed by Rudolf Gantenbrink.

Figure 22 (above)
Inside the pyramid.
The walls are bare and without script or ormanent. (The metal struts are modern.)

Figure 23 (left)
At the end of the Great Gallery, the ways divide. The lower shaft blocked by a grille, leads to the Queens' Chamber.

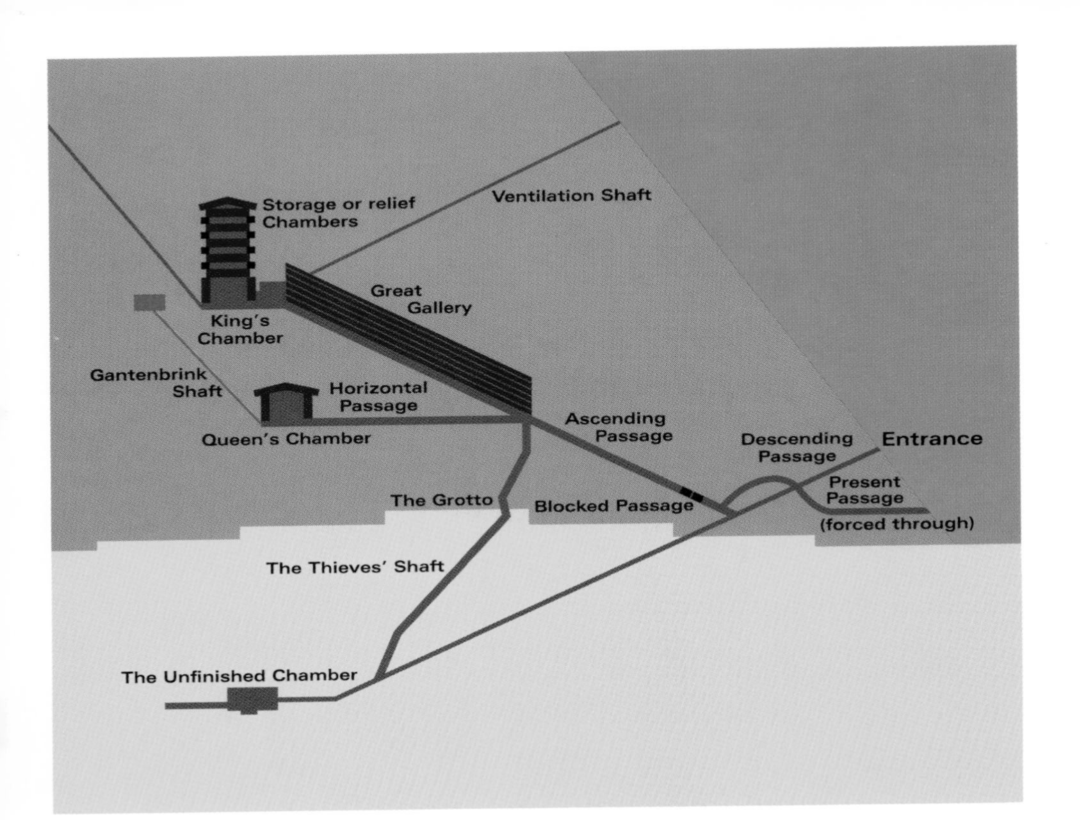
Storage or relief
Chambers
Ventilation Shaft
Great
Gallery
King's
Chamber
Gantenbrink
Shaft
Horizontal
Passage
Queen's Chamber
Ascending
Passage
Descending
Passage
Entrance
Present
Passage
(forced through)
The Grotto
Blocked Passage
The Thieves' Shaft
The Unfinished Chamber

Figure 24 (opposite, top)
Sketch of the pyramid showing the three chambers. In spite of its dimensions, the Great Gallery is not counted as a chamber. The Gantenbrink shaft is shown in red.

Figure 25 (opposite, bottom)
Upuaut on its way!

Figure 26 (above)
The mysterious door at the end of the Gantenbrink Shaft.
Of the two metal clasps, the left has broken off. At the lower right-hand corner, a triangular portion is missing.

Figure 27 (right)
Scientists from Waseda University, Tokyo, x-rayed the pyramid by means of modern electronics. This is the title page of their study.

Figure 28 (overleaf)
Is the sphinx 7,000 years old?

Studies in Egyptian Culture
No. 6

Non-Destructive Pyramid Investigation (1)
—By Electromagnetic Wave Method—

by
Sakuji YOSHIMURA
(Egyptian Archaeology)
Shioji TONOUCHI
(Geophigics, Dr.)
Takeshi NAKAGAWA
(Architecture, Dr.)
Kazuaki SEKI
(Egyptian Architecture)

1987
WASEDA UNIVERSITY
TOKYO-JAPAN

Figure 29 (opposite)
The Great Gallery

Figure 30 (above)
Two metres in front of the door lies the broken-off piece of metal. (Computer reconstruction)

Figure 31 (overleaf)
Rudolf Gantenbrink

Susa. Today it can be seen it in the Paris Louvre. The *Codex Hammurabi* consists of 282 paragraphs; according to Hammurabi these were given to him by the god of heaven – in just the same way as Moses received the tablets of the Commandments directly from the hand of God. In the 'foreword' to his rule-collection, Hammurabi expressly says that 'Bel, the Lord of heaven and earth' had chosen him to 'spread justice through the land, to destroy the wicked and to prevent the strong suppressing the weak.'[29] And of course the people looked forward to the return of their law-giver.

All that we can, in retrospect, be sure of is that Hammurabi achieved *something* remarkable, and distinguished himself from all his contemporaries through various unusual deeds. Of course it would be possible to assume that divine origin was attributed to him only *after* his death – if it were not for the stone pillar which bears his own testimony, written during his life, that he had been chosen by the gods. Should we call the supreme law-giver a supreme liar? That would be like accusing Moses of making up the story of receiving the stone tablets on the holy mountain.

We clever and superior modern people 'know', of course, that the seed of King Hammurabi could not possibly have come from the sun-god. But how do we know it? We were not there, and the skeleton of Hammurabi has never been subjected to genetic research. It is fairly typical of human logic that we reject Hammurabi's claim to having contact with other-worldly beings, while accepting the stories of Moses and other prophets.

The Assyrian king Assurbanipal (668–622 BC), in whose clay-tablet library the *Epic of Gilgamesh* was discovered, was also immaculately conceived. He was the son of the goddess Ishtar, who suckled him as a baby. Ishtar must have come from other worlds because it says in a cuneiform text: 'Her four breasts lay upon your mouth; you suckled upon two, in two you hid your face.'[30] That's right, *four* breasts – enough to make some of us envious. This King Assurbanipal received the authority for his decisions from the 'divine advice' of the gods Bel, Marduk and Nabu. The last was the omniscient god from whom humanity learnt to write in script. In the Louvre there is a cylindrical relief sculpture upon which Nabu is depicted next to Marduk. Nabu's chief temple was situated in Borsippa and bore the name 'Temple of the Seven Command-Transmitters of the Heaven and the Earth' – strange name.

Was all this just self-aggrandizement on the part of the ruling elite? Did their authority come to depend on the people and priests believing that they were of divine origin? Personally I do not think so. Not every king and founder of religions laid claim to bearing a 'heavenly seed' within him; only some from those undatable mists of time were convinced that they had a quite specific genetic code to pass on. We must not forget that similar stories appear in many different traditions and in various undatable texts – the Egyptians, Enoch, the Jains, and of course the Apocrypha of the Old Testament! The last also speaks of divine teachers, even if they are called 'fallen angels'; and there, as well, in the hidden mists of Jewish tradition, we find an abundance of characters whose seed was not of earthly origin. Of course these things do not find a very receptive audience; people hold them at arm's length. And suddenly Erich von Däniken is said to be in cahoots with a bunch of idiotic racists, as if it was I who had invented the idea of 'heavenly seed' and 'chosen ones'. I cannot be held responsible for such concepts – they come straight out of ancient traditions and texts which are holy for many peoples.

So Noah, the survivor of the flood, for example, was not just anyone. His earthly father is named as Lamech, but in fact Lamech was not his physical father – everyone can read this for himself in the Dead Sea Scrolls.[31] It says there that one day Lamech returned home from a journey which had taken more than nine months. Once home he found a baby who did not belong to his family – it had different eyes, different hair-colour and a different kind of skin. Furious, Lamech went to his wife, who swore by all that was holy that she had not slept with a stranger, let alone a soldier or a son of heaven. Worried, Lamech went off to ask his father's advice. This was no other than Methuselah. He had no light to shed on the matter, and so, in his turn, went off to ask *his* father, Lamech's grandfather. And who was that? Our friend Enoch. He said to his son Methuselah that Lamech should accept the boy as his own son and not be angry with his wife, for the 'guardians of the sky' had placed the seed in his wife's womb. They had done this so that the cuckoo's egg, as it were, should grow into the progenitor of a new race after the flood.

This episode shows that Enoch – who later travelled up to the clouds in a fiery chariot – was already informed about the approaching catastrophic deluge. Who had told him? The 'guardians of the sky'. And who arranged the artificial fertilization of Lamech's wife? The same space travellers.

It is with such examples that I try to illuminate accounts and traditions that are to be found the whole world over, and that have existed for many thousands of years. This divine jet set, these innumerable sons of gods, come leaping out at one from almost every mythology in the world.

Gods of Yesterday – Gods of Tomorrow

The culture of the Tibetans, which grew to greatness in lofty valleys cut off from the rest of the world, is familiar with the 'highest king of heaven' or the 'holy one above'.[32] The Tibetans distinguish between a transcendental heaven and the firmament.

> The oldest Tibetan kings were called 'heavenly thrones'. They descended from the heavens in the service of the gods, and returned once their rule had ended, without passing through death.

They possessed unimaginable weapons, with which they either destroyed or controlled their enemies. The appearance of some of these weapons has been preserved in folk memory – the 'thunder-hammer', for instance, which is still honoured today in Tibetan temples. There must be more to this than mere fantasy; these 'thunder-hammers' are a reality, even if we cannot imagine how they worked.

The legend about the great Tibetan king Gesar tells that he was assumed by a 'heavenly apparition of light'. Once he had created order in the land he vanished back to his home in the sky, naturally promising to return one day. Like the mysterious primeval rulers of China or the god-kings of ancient Egypt, King Gesar was a teacher of mankind. Like them he was thought of as a 'human-maker', before whose coming human beings still lived like animals. In the royal genealogy of Tibet, the so-called *Gyelrap*, 27 kings are recorded; 7 of them descended a ladder from the firmament to earth. And even the oldest texts themselves flew down to earth in a box. The great Tibetan teacher with a tongue-twister of a name, Padmasambhava (also known as U-Rgyan Pad-Ma), brought indecipherable texts from the heavens to earth. Before his departure, his pupils deposited these texts in a cave to preserve them for 'a time which would understand them'.[33] The same teacher vanished in front of his disciples' eyes and returned to the clouds. He was not, apparently, 'beamed up', but 'a horse of gold and silver appeared', and everyone watched as he took off into the clouds upon this steed. Ring any bells? Enoch and his steed might well have been a close acquaintance!

I am almost embarrassed to add that the holy books of Tibet also make reference to impossible numbers. Four great divine kings are recorded, each of whose life-span amounted to nine million earthly years. Also described are various cosmic dwelling places, reached after long journeys through space. The numbers and periods mentioned remind one strongly of Einstein's theory of relativity – a major difference being, of course, that the Tibetan books *Kandshur* and *Tandshur* are thousands of years old.[34]

But it is not only in the Near and Far East that such ideas were prevalent. In America, the native Indians had very similar legends. The stories of the Wabanaki tribe refer to the teacher Gluskabe, who taught them fishing, hunting, hut-building, weapon-making, medicine, chemistry and of course also astronomy. Before he finished his work on earth and took off for the stars, he promised to return in a far-distant, future time.[35] What a surprise!

I have talked about the Maya god Kukulkan in another book.[36] In passing I will just cite one quote: 'The people have the firm conviction that he travelled up to the heavens.'[37] And for anyone who has not guessed it – he also promised to return.

These fragments of folk memory and religion do not need a Sherlock Holmes to connect them. And personally, I think it is a load of nonsense to say that various peoples all over the globe had taken their expectations of the gods' return from Christian missionaries. What came first for heaven's sake – the Christian or the other texts?

Whichever culture one examines – and there are many more I have not mentioned (such as the Aborigines in Australia, the Chinese, the Incas: remember that the Christian conquerors Pizarro in Peru and Cortes in Mexico were greeted as gods who had returned) – similar or nearly identical legends are to be found. Gods with their return tickets are a world-wide phenomenon, and the examples I have quoted in this chapter are no more than the tip of the iceberg.

Who Will Return?

But *who* is supposed to return and *when*? Christians and Jews await the Messiah, Muslims the Mahdi – which is actually just another name for a messiah figure. The word 'messiah' originally meant 'the anointed one'. It comes from the Hebrew *maschiach* (*christos* in Greek), meaning the anointed king; but an earthly king cannot be meant, for, as the famous professor, Dr Hugo Gressmann, wrote, the term 'messiah' precludes the idea of a human being: 'Messiah is the name for divine being, one who is also thought to have existed before there were human beings.'[38]

Let us look at the common denominator of all these 'messiah' concepts.

- He has great power.
- He brings a new order.
- He is justice personified.
- He is inspired, elected and directed by God.

According to different religions he is:

- a 'son of man' conceived by the divine (seed, embryo, *karma* of the divine), who has usually dwelt for a while on earth, is

then assumed into the heavens, and will one day return
- either one or many extraterrestrial, god-like beings, who once came to dwell on earth

In many traditions, the return of the gods is associated with some kind of day of judgement or final reckoning, and with catastrophic natural events. Every religion adds its own colour and interpretation, twists the story a little or a lot to reinforce its own message and assure the salvation only of those who believe in it. But the legends at the core of all these beliefs are far older than particular religions, whether Christian, Muslim, Jewish or Buddhist. So let me repeat: who is going to come? Whose judgement should we fear? Who will return with heavenly armies and mighty turbulence in the firmament?

Palaeo-seti philosophy can offer to these questions an answer which accords with all traditions. It is a theory which confirms many texts and solves many separate riddles. But in contrast to religions, palaeo-seti philosophy does not require any faith or belief – just unprejudiced, rational examination of its ideas and propositions; for it is, unlike the messianic expectations of religion, based on logic and reason.

Goodbye Papa!

The alien space travellers who dwelt upon the earth thousands of years ago and gave the human race a genetic lift-off – the same space travellers that are recorded in ancient literature as gods, angels, fallen angels etc – departed at some point. A few privileged people were allowed to depart with them; they too took their farewells. What was said to those who remained behind, those who would probably have liked to go on such a journey themselves? Here is an imagined farewell dialogue between Enoch and his son Methuselah:

Enoch: It is time my son. They will come at dawn to fetch me.
Methuselah: Father, will we see you again?
Enoch: No. At least, your generation will not. I was told that during my absence several millennia will pass on the earth.
Methuselah: How can that be? Surely death comes to all?

Enoch: True. But other laws of time are at work in the cosmos. When the guardians return after thousands of years, the earth and human beings will have changed.
Methuselah: This is beyond my understanding. But this is what the guardians have told you. And where will you travel to?
Enoch: Do you see the bright stars in the belt of Orion? Now extend that line by 6 feet. There you will see a small star, not so bright, of a yellowish colour. That is the home-sun of the guardians. There is an earth more beautiful than ours. That is where I am going.
Methuselah: Father, you have been chosen to journey to heaven as a living man – I envy you.
Enoch: No, my son, I am not going to heaven. The heaven which men long for is a place of absolute happiness. We can only reach such a heaven after death. I am journeying into the cosmos.
Methuselah: I do not understand the difference between heaven and what you call 'cosmos'. Look up at the glory of the heavens; up there is peace and beauty. The guardians can travel there upon their fiery steeds. Their power is boundless. To us it seems that they are immortal. It must be the same as heaven, even if you call it 'cosmos'.
Enoch: The time of my departure draws close. Do you hear the commotion of the people? They are gathering to hear my farewell words. The guardians have warned me not to allow anyone to approach the place where the fiery steed descends. The same applies to you and your family. And now, son Methuselah, I have explained everything to you and given all the books I have written into your safe-keeping. Preserve them. Have them copied many times – and make sure that not a word is altered. Even if you and your sons and grandsons do not understand my words, later generations will, and will be grateful that you altered nothing. The guardians have told me that these books should not remain secret. Therefore give them to the future generations of the world.

However hard Enoch may have tried to make his listeners understand that he was travelling into space and not into heaven, following generations would not have grasped the difference. In later times, those who had themselves not witnessed the visit of the 'gods' would have found little meaning in the texts they read. These beings who had descended in their great-great-grandfather's time *must* have been divine messengers of God – and so the idea of angels came about. It is human nature to look for sense – to the extent of making up nonsense.

The thinkers and philosophers of each generation, the 'men of wisdom', would – just as in the imagined scenario of the Sacred Berlitz Stone – have subtly altered the texts to make them more understandable. Such wisdom might well think that a passage describing a strange steed that shone, thundered, had four feet and flew should be altered to make clear that a flying horse was meant. Alien beings could easily be called angels, the commander could easily become 'highest', and descriptions of the inside of a spaceship could be seen as the dwelling of angels and the throne of God. In the following comparison of the present Book of Enoch with an imagined original source, I will try to expose this interpretative process.

Imagined source:

This was my experience: first I saw clouds, and then, as we were lifted still higher, I noticed a mist that grew ever thinner and finer. And suddenly we were amongst the stars, yet something also flashed like lightning around us. I was so cramped that they had to lift me from the chair. I walked along a passage until I drew near to a wall which seemed to be composed of glittering stones. I also noticed reddish points of light that flashed up and down this wall. Then I stepped into the starship. Inside was as bright and shining as outside, but now the floor was made of tiles, under which shimmered a faint light. The roof was the most beautiful: as though through a transparent dome I saw the starry sky; and guardians who kept arriving and leaving in smaller flying machines, and undertook all kinds of work. Then we had to embark once more, entering a larger starship. Inside it all doors were open, but I saw indescribable configurations of light in front of each door. The guardians explained that these were sensors and door-shields. The centre of the starship was enormous and indescribable. In the very middle, upon a raised platform, stood a seat; and all around, dully shining, a great circle of glass. Upon it I saw a shining sun, and many guardians who were working outside the ship. Upon the seat sat the commander, dressed in a snow-white tunic. I threw myself in front of him on the floor; but he came towards me, spoke words of greeting to me and said: 'So it is you who have the task of spreading order and justice down below?'

From the Book of Enoch (14:8ff; 71:11ff) as it is today:

This appeared unto me: behold, clouds invited me upwards, and a mist drew me on; the course of the stars, and lightning, drove and pressed me, and winds gave me wings and lifted me into the heights. They bore me into heaven. I entered there and drew near to a wall built of crystal stones and surrounded by tongues of flame; and this began to make me afeared. I entered into the tongues of flame and drew close to a great house built of crystal stones. The walls of that house were like unto a floor tiled with crystal stones, and the ground also was of crystal. The roof thereof was like unto the course of the stars shot through with lightning, and fiery cherubim went between...and behold, there was another house, greater than that one; all its doorways stood open before me, and it was built of tongues of fire. Its glory and greatness were in all ways such that I cannot describe it unto you ... I looked thither and descried within a high throne. The shape of that place was of a bracelet; all around it was something like unto the shining sun, and which had the appearance of cherubs ... The great Majesty sat there; His dress was more radiant than the sun and whiter than driven snow ... Then did I fall upon my face; my whole body melted and my spirit was transformed ... He drew close unto me, greeted me with His voice and spake: Thou art the one, which art born to righteousness.

Exegesis Through the Ages

What a drama when space travellers become angels and cherubim, when officers become archangels, and a commander is made into the 'highest' or – God help us! – into God. What chaos, when simple electric discharges are made into fiery tongues, and a command bridge is transformed into an indescribable glory! It is of course understandable that the commander's seat is made into a high throne, and the commander himself becomes a great Majesty. It is comforting, at least, that our loving God does not slip in through the back door into this piece of text. That would, of course, have been rather unsuitable anyway, since 'he drew close to me, greeted me with his voice ...'. Gods do not usually deign to shake hands with an earthly visitor – that would have been too much even for the exegetes; so they left it as the great Majesty.

The extraterrestrial visitors of Enoch's time were familiar with gigantic interstellar distances. They knew that a journey home and then back to our solar system would take a few thousand years. How could they get human beings to understand this? They would have pointed to the starry sky and said: 'We are going now, but we will return. Write it down in your books, pass on the message to your descendants; all future generations should remember that we will return!' And when human beings asked them when they would return – in months, years or millennia – the ETs themselves would not have known a precise answer. They might well have replied: 'Some time or other we will come back. Be prepared for our return, remember the commandments we have given you, so that we do not have to destroy the human race all over again.'

And if people asked them what recognizable signs would accompany their return, they might well have pointed to the moon and the stars and answered: 'For those upon the night-half of the globe, it will seem that the moon grows dark, that shining stars fall to the earth. For those upon the globe's day-lit half, it will seem that golden mountains crash down from the heavens. The people who are prepared for our return, who expect us, who understand the signs in the sky, will be full of joy. They will dance and rejoice because we will bring a new, more just, order to the earth. But those who have adulterated the texts, who have forced their fellow human beings to believe their versions of the truth, will be overcome with panic. They will be frightened of us and of their own followers. They will hide themselves and call upon their false gods. But it will be in vain, for there are no gods.'

But of course the extraterrestrials were aware that the texts would be tampered with and reinterpreted through the ages. For that reason they left their traces in many different parts of the world, made sure that many different human societies on earth would have a written record of their coming. At some time in the future, global communication would make possible a mutual exchange of these traditions. And then, they hoped, the core of truth at the heart of all these different accounts would emerge. People would have to begin to make comparisons. One and one makes two.

Palaeo-seti philosophy actually turns received wisdom – which generally manifests in one of two opposing ways – on its head. There are basically two kinds of people: the believers and the non-believers. Each group has been differently educated and imbued with different values; but they are in agreement on one thing – the human being is the only intelligent life-form in the universe. The believers think God created the earth in a (symbolic) act of six days, and then rested on the seventh. After God made the plants and animals, he formed man as the crowning glory of creation. Hallelujah! The unbelievers, on the other hand, hold fast to the theory of evolution. In a process lasting millions of years, amino acids formed cells, then simple life-forms, then more complicated life-forms, until – as the high-point of evolution – *Homo sapiens* appeared. We are the summit of evolution. Hallelujah once again!

In both cases we are seen as the highest form of life – unique in the universe. What should we want extraterrestrials for, even if all the holy books in the world provide us with evidence of their existence?

Upsetting Old Values

And here they come! All sorts of different spaceships: multi-storeyed, flat, golden and copper-coloured, smaller *vimanas* and gigantic structures that look like towns placed one upon another. They pass across the full moon and cause tumultuous disturbances to our oceans. Humanity is terrified, shocked, astonished. *That* was not what we were expecting, neither the believers nor the non-believers. The Christians will run into their churches and ask the priests: 'Has the day of judgement come?' Muslims will pray to Allah, devoutly wishing that it is the Mahdi who has returned, and that he has at last come to sort out the unfaithful after long ages of waiting. The Jews will go to their synagogues and waylay their chief rabbis, and all Jerusalem will be awash with people, since tradition has taught that the Messiah would descend there. Only the scientists will shake their heads in dismay as they get out their sensors and telescopes and comb the skies, finally having to accept the fact that extraterrestrials have taken up their positions all around the globe.

But the believers, who hold fast to the idea of their own messiah, will lose their grasp of reality; they will not be able to relate their entrenched belief-systems to these new developments. They will be too inflexible to come to terms with the new (and at the same time, ancient) facts which confront them. They will be incapable of altering their ideas to encompass a new global politics and universal religion. So they will become unbelievers – of reality. They will see the extraterrestrials as messengers of the devil, who have only appeared on the horizon in order to shake their faith. They will be bitter and confused, because they are unable to accept the evidence in front of them; and ultimately they will die because they do not understand anything any longer.

The real believers on the other hand – those who can live with the facts that now present themselves, who no longer need any kind of faith because they *know* – will blossom. Until then, all knowledge was built on the past; to this will now be added a knowledge that comes from the future, the knowledge and know-how of the ETs, who have already overcome the problems which plague us. For them, our future is already the past. Humanity will rush to draw upon their knowledge, like bees sucking honey. 'How did you solve your environmental problems? How did you deal with the dangers of a population explosion? What kind of religion do you have, and upon what is it based? How are your spaceships powered and how does interstellar radio work? How does one heal a cancer tumor and how can life be prolonged? What political system is the most just and how do you punish your criminals?' Thus will we leave the one-way road of knowledge behind us and join the sliproad onto the eight-lane highway. When the universe opens its doors to us, a truly *heavenly* epoch will begin. But only for the believers – sorry, I mean those who can cope with reality.

This reversal of values, this new philosophical approach to the 'second coming', is within sight. Religions will struggle against it and denounce me as a heretic; they will call me a tempter and pseudo-prophet, refusing to accept that they are the ones who helped keep alive the expectation of the gods' return for thousands of years; that they carved away at their own messiah image – or whatever they like to call their saviour – until it fitted nicely into their glass cabinets, like a museum piece. All the glass cabinets of other religions were, of course, considered fit only for smashing. Each religion asserted that its own teachings were superior to all others. I, for one, have never got involved in this one-upmanship. The cap didn't fit, and I couldn't see the point.

The Seeds Bear Fruit

We know very little about the real power and genetic technology of extraterrestrials. But they must, at the very least, be a few thousand years ahead of our capabilities, otherwise they – or their ancestors – could not have visited us in the ancient past. The history of modern technology and science teaches us that everything is made continually more perfect, small and effective. Computer technology demonstrates this with more and more microscopic chips, thousands of millions of bytes and continually increasing operation speeds.

For example, in the middle of the 1980s already, any better-quality PC could manage a computation speed of several megaflops (flops = floating point operations per second; megaflops = a million flops). Large computers like the Cray 2 attained gigaflops (a thousand million flops) by the beginning of the 1990s. One year later, ten gigaflops were achieved; and while I type these lines, the 100-gigaflop computer has been announced – the CM 5. The teraflops computer (1 billion flops) is already being developed and people are starting seriously to discuss the possibility of a ten-teraflop computer.

That may well be termed a dizzy speed of progress. But what are ten short years compared with millennia of evolution? A speck in the ocean. What will computers be up to in 50 years' time? They will think for themselves, programme themselves and hold conversations with us. They will be capable of instantaneous, impeccable translation of any language of the world into any other. There will be law computers that will be able to judge a case faster, better and more justly than human beings. Computers will build computers; and instead of the TV screen in your living room, there will be a three-dimensional hologram projection.

In the field of genetics, things have been achieved that biologists of the old school would not even have dared dream about. Within the next 20 years, geneticists will be able – either at the embryo stage, or even before conception – to prevent parents passing on inherited illnesses to their offspring. As long as our laws and ethical codes allow it, they will be able to construct human beings with quite specific characteristics – real works of genetic art. People say that this is 'playing at God'; but they forget that the God (or better, the gods) of the Old Testament created man 'in his image'. He programmed him in the way he wished, and clearly also kept tinkering with his descendants. I hope it is clear by now that this 'God' cannot be the creator of the universe. The geneticists who 'play at God' are no more synonomous with creation and the spirit of the universe than were the 'gods' of the mythologies. A computer may seem divine to an ape – but this does not make it divine.

If such advances can happen in the short span of 50 years, what can be achieved in several thousand years' worth of scientific and technological development? How far have the ETs advanced by now? If they could already predetermine the genetic characteristics of a foetus millennia ago, what are they now capable of? Can they perhaps influence the genetic code from a distance, by means of some invisible ray or beam? Can they gain access to our brains? Did they perhaps insert a code into our genetic make-up thousands of years ago, so that after a predetermined number of generations certain messages would be released into the brain? Is it possible that we contain coded messages and information that are woken by specific stimuli, so that we start to become conscious of them?

Every modern geneticist is familiar with the so-called genetic 'junk'. This refers to apparently useless and pointless sections of DNA (deoxyribonucleic acid). They seem purposeless because they have neither a proper beginning nor a proper end. Inherited characteristics are usually 'plugged' with a kind of stopper, which only fits onto the right counterpart material. Dr Beda Stadler, professor of genetics at Bern University, compares this with Lego bricks. Our DNA contains approximately 110,000 active genes, between which can be found many fragments of genetic junk. Is it really junk? Or does it have a quite specific task which had eluded geneticists up to now? It is hard to believe that thousands of years of evolution have carried with them so many useless fragments of genetic rubbish.

Even though our knowledge is continually pushing forward its frontiers, we still know next to nothing about its universal context. Yet we continue to behave as though we know it all. I am not in the least put out by the prophets of the Jain religion, the *tirthamkaras*, nor by the 'super-Buddha'. My theories are not in the least undermined by such phenomena, nor by such people as the living Sai Baba in India, who works miracles. Is it not possible that the coded message has just been activated in him a little too early? We know from experience that human genes only release particular messages after a certain period. A six-year-old boy does not grow a beard, nor is he sexually mature. Body hair and sexual maturity only come about when certain physical stages have been reached; particular hormones are then activated and released by genetic codes and messages. But the code for body hair was present from the beginning – was slumbering in the infant already, was programmed into every cell at the time of conception even. The message was there, but the time for it had not yet arrived.

Is it not possible that 'genetic junk' fulfils the same function in us? Do we carry within us information that requires only a signal – some message or other – to awaken it? Computer technology is already experimenting with atomic switches which introduce single electrons to activate the binary process of 'yes' or 'no'. This astonishing switch – fast as the speed of light – was discovered by the Russian physicist Konstantin Licharev. It is called the single-electron tunnelling effect (SET); its efficacy has since been proved, and it is thought to be a model for the development of ultimate miniaturization in micro-electronics.[39] But if an electron can serve as a switch to direct a computer in one way or another, it can surely also awaken an innate genetic code or message.

Return in Other Forms

Palaeo-seti philosophy interprets the idea of the return of the gods as a return of those extraterrestrials who long ago visited our ancestors. To alleviate the shock of this return, prophets are released into humanity to prepare it. Such prophets may have received their knowledge in one of several ways.

- They may themselves be extraterrestrials in human guise.
- They may be human beings whose embryos were externally programmed ('sons of man').
- All of humanity bears coded genetic messages which are only released when certain conditions are fulfilled (as in the growth of body hair); this would occur at different times in different individuals.
- Perhaps the whole of humanity bears this genetic information within itself, but it is only released in certain individuals by means of some activating beam or ray directed from outside the earth (as in the electron switch).
- Perhaps only certain individuals bear this extraterrestrial message within themselves.
- Such genetic information may only be beamed into single individuals when the ETs think the time is ripe for it.

I think the fifth possibility the least likely, since we all ultimately descend from the same stock, regardless of whether we mean the symbolic Adam and Eve or our post-diluvian ancestors. The sixth option is not impossible, but still highly speculative.

In the Book of Enoch (39:1) we read:

> In these days will descend from the high heavens the elect and holy children, and their race will be joined with that of the children of men.

Is Enoch pointing us to the second possibility in the above list? If so, how did he know? From the 'guardians of the sky'? From whom else? And how do the prophets come to describe those utopian scenes to us in their ancient books? In the Revelation of St John, we read (9:1–3; 7:9–10):

> And the fifth angel sounded, and I saw a star fall from heaven unto the earth: and to him was given the key of the bottomless pit ... And there came out of the smoke locusts upon the earth ... And the shapes of the locusts were like unto horses prepared unto battle ... And they had breastplates, as it were breastplates of iron; and the sound of their wings was as the sound of chariots of many horses running to battle. And they had tails like unto scorpions, and there were stings in their tails: and their power was to hurt men five months.

And three chapters later, in 12:7–9:

> And there was war in heaven: Michael and his angels fought against the dragon; and the dragon fought and his angels, And prevailed not; neither was their place found any more in heaven. And the great dragon was cast out, that old serpent, called the Devil and Satan, which deceiveth the whole world: he was cast out into the earth, and his angels were cast out with him.

This book was supposed to have been written by the holy apostle John, but every researcher knows this is not true. The 'secret revelation' does not come from John, but from some editorial team working in the years between AD 90 and 100. Of course they did not just make the text up out of the blue – they were working from older texts. One can find similar descriptions in the Apocrypha, mainly (but not solely) in the Book of Enoch; also in short sections of the Old Testament, in the Book of Daniel for example. This all suggests that there must once have existed an older, original, common source. Someone must originally have written the first text, must also have experienced these frightening visions. Or must he?

I do not dabble much in psychology; I do not value it very highly, and I know that it tells one either everything or nothing – depending on whether one believes in it or not. The following approach seems to me much closer to the truth.

We have all seen films like *Star Wars* and *Star Trek*. We know the special effects which can be achieved nowadays in modern films. I imagine that the extraterrestrials have a far more advanced 'vision-technology'. Perhaps they show their films in 3D, without the need for virtual reality headsets. Cinematic technology involving laser holography would conjure perfect illusions.

Now the 'guardians of the sky' had very close connections with Enoch. At the end of their sojourn on earth they even took him with them on their great journey. Why should the ETs not have shown films to some of their favourite earthlings? Human descriptions of these films would easily have turned warring robots into 'locusts ... like unto horses ... and they had breastplates', and 'the sound of their wings was as the sound of chariots'. And the poor archangel Michael – who of course had no such name in the ETs' film, but was given it by later interpreters – 'fought against the dragon'; and at the end of all this, one side wins and the loser is thrown into the depths – as in any film of that sort.

Someone wrote it down; it may well have seemed to him like a vision. Later generations certainly made a 'vision' out of it; and finally, various fragments of this apparent vision ended up in the writings of various prophets. Later on, a group of editors cobbled together the apocalypse and the 'secret revelation', and even made venerable old John responsible for it.

Not all that is to be found in the holy texts has to be vision and revelation. The most likely explanation is often the most banal. All that is needed is to be prepared to look at things from a different point of view.

NOTES

1 Stearn, J, *Edgar Cayce: Sleeping Prophet*, Bantam, 1983; Church, H W, *Die 17 Leben des Edgar Cayce*, Geneva, 1988

2 Sandweiss, S, *Sai Baba, the Holy Man and the Psychiatrist*, M G Singh, 1975

3 Ihlan, O, 'Wunder sind mein Wesen', in *Der Spiegel*, No 38, 1933

4 Eggenstein, K, *Unknown Prophet Jakob Lorber*, Valkyrie Publishing House, 1979
5 Mirza Mubarak Ahmad, *Der verheissene Messias*, 1977
6 Charon, E, *Der Geist der Materie*, Vienna/Hamburg, 1979
7 Baumgartner, W, *Hebräisches Schulbuch*, Basel, 1971
8 Küppers, W, *Das Messiasbild der spätjüdischen Apokalyptik*, Bern, 1933
9 Klausner, J, *Der jüdische und der christliche Messias*, Zurich, 1943
10 Dürr, L, *Ursprung und Ausbau der israelitisch-jüdischen Heilandserwartung*, Berlin, 1925
11 Landmann, L, *Messianism in the Talmudic Era*, New York, 1979
12 Schomerns, H W, *Indische Erlösungslehren*, Leipzig, 1919
13 Ayoub, M, *Redemptive Suffering in Islam*, New York/Paris, 1978
14 Dalberg, F von, *Scheik Mohammed Fani's Dabistan oder von der Religion der ältesten Parsen*, Aschaffenburg, 1809
15 Widengren, G, *Hochgottglaube im alten Iran*, Uppsala/Leipzig, 1938; Reitzenstein, R, *Das iranische Erlösungsmysterium*, Bonn, 1921
16 Abegg, E, *Der Messiasglaube in Indien und Iran*, Berlin/Leipzig, 1928
17 Schomerns, H W, op cit
18 Roy, D P, *The Mahabharata, Drona Parva*, Calcutta, 1888
19 Däniken, E von, *Der Götter-Schock*, Munich, 1992
20 Däniken, E von, *Prophet der Vergangenheit*, Düsseldorf, 1979
21 Karst, J, *Eusebius-Werke, Vol 5: Die Chronik*, Leipzig, 1911
22 Wahrmund, A, *Diodor von Sicilien: Geschichts-Bibliothek*, Book 1, Stuttgart, 1866
23 Roth, R, 'Der Mythos von den fünf Menschengeschlechtern bei Hesiod', in *Verzeichnis der Doktoren, die Philosphische Fakultät*, Tübingen, 1860
24 Glasenapp, H von, *Der Jainismus: Eine indische Erlösungsreligion*, Berlin, 1925
25 Charon, E, op cit
26 Däniken, E von, 'Embryo transfer in ancient India', in *Ancient Skies*, No 3, 1991
27 Glasenapp, H von, op cit
28 Jeremias, A, *Handbuch der Altorientalischen Geisteskultur*, Berlin/Leipzig, 1929
29 Frischauer, P, *Es steht geschrieben*, Zurich, 1967
30 Jeremias,A, op cit
31 Burrows, M, *Mehr Klarheit über die Schriftrollen*, Munich, 1958
32 Hermanns, M, *Schamanen, Pseudoschamanen, Erlöser und Heilbringer*, Wiesbaden, 1970

33 Grünwedel, A, *Mythologie des Buddhismus in Tibet und in der Mongolei*, Leipzig, 1900
34 Feer, L, *Annales du Musée Guimet: Extraits du Kandjour*, Paris, 1883
35 Breysig, K, *Die Entstehung des Gottesgedankens und der Heilbringer*, Berlin, 1905
36 Däniken, E von, *Der Tag an dem die Götter kamen. 11. August 3114 v. Chr.*, Munich, 1984
37 Breysig, K, op cit
38 Gressmann, H, *Der Messias*, Göttingen, 1929
39 Schön, G, 'Die kleinsten elektronischen Schalter – Cluster aus 55 Goldatomen', in *Spektrum der Wissenschaft*, April 1994

4

Tracking Down the Truth

'Mockery ends where understanding begins.'
(Marie von Ebner-Eschenbach, 1830–1916)

Where are the traces of the extraterrestrials? Everywhere. Most people fail to see them: there is only circumstantial evidence, and nothing is proven. But whoever does not see the traces and signs of them in the great legends and mythologies of the world must be half-blind. Since nearly everyone seems to have this vision deficiency, one may ask why these ETs did not leave more obvious markers and indicators of their visit here. What good are religious texts and tales from ancient times, what use are strange traditions of 'impossible numbers' if everyone can make of them what they will?

Proof is needed that cannot be refuted. Only then will science sit up and take notice. Is that so? But how often in the past were scientific proofs demonstrated and then rejected because they did not fit in with a religious world-view? How often has one branch of science proved something, only to have it refuted by another branch that did not like the sound of it? How often have irrefutable scientific proofs – yes, this has happened! –

been generally undermined for ideological reasons? The geneticists in every laboratory could recount epics on such a theme! They can prove with ease how reasonable, important and forward-looking genetic research is. And how does the media respond? Hands off! Dangerous! Dreadful! Should be outlawed immediately! What did Albert Einstein say? 'Two things are infinite: the universe and human stupidity' (although he was still unsure about the former).

So what kind of irrefutable proof could the extraterrestrials have left behind? Some kind of sculptures on cliffs or mountains? No. Over thousands of years they would crumble and erode. Might they have put up some kind of buildings, say pyramids? No, for the same reason as above. If bacteria, termites or self-righteous people do not destroy such buildings, earthquakes, floods, volcanic eruptions and other natural catastrophes would do the job.

But could they perhaps have left an indestructible text somewhere? Could they? Where then? In which building, inside which mountain? No, for the same reason as before.

But why is a building necessary? The ETs could have left evidence behind in a metal form, or in some artificial material – at any rate something that would stand the test of time. There are actually such remains; but unfortunately religion forbids their scientific investigation.[1] And from which indestructible metal would the 'tablets of the gods' be made? Silver, gold, platinum? All of them can be melted down. Steel then? Super-steel? But where are the thick tank-plates from the First World War? They have all rusted! And what about the remains of the thousands of aeroplanes which were shot down in the Second World War? That was only yesterday, comparatively speaking! Even the few remains that are preserved in museums will have disintegrated in a thousand years' time.

But the 'guardians of the sky' must have left waste materials lying around; surely these could be found? No, it would be absurd after such a long time to expect to discover objects that had been thrown away. Nature has absorbed them.

But there must be some way of transporting messages from the past to the future. I agree. But for this to happen, two conditions have to be fulfilled:

- The message must be in an indestructible form.
- The message must never fall into the hands of the wrong generation.

Who is the wrong generation? All those who are incapable of properly evaluating such information. They would wreck the message without deciphering it. If it was in the form of higher mathematics, only a mathematically highly advanced society could decipher it. If it consisted of microfilm, only a society could understand it which can read microfilm. If it was coded in computer language, only those who had advanced knowledge of computer technology would make sense of it. If the message was left in the sterile environment of the moon, the (almost) sterile environment of Mars, or perhaps in a satellite orbiting the earth, it would only be discovered by a society that had begun to voyage into space. And if the message was concealed in genetic material, only a society capable of deciphering DNA would come upon it.

But in order for a society even to think of looking for such a message, signs and traces must be left, indicators that stimulate a search. What I do not know about does not make me sweat.

The Message of the Gene

It would seem likely, at the present stage of palaeo-seti research, that the extraterrestrials' message may have been implanted into human genes and also into those of specific forms of vegetation. The ETs of thousands of years ago relied upon human, or rather scientific, curiosity. 'The gods created man in their image', says tradition. They did not only create man, however, but also, according to legend, exquisite and unique plant-forms. All that the extraterrestrials needed to do was to implant a few gene-sequences (modifying the DNA – also called 'artificial mutation') into the human genome and certain 'divine plants'. Curiosity is an expression of intelligence, which is a quality that became characteristic of the human race after this artificial mutation had taken place. All our knowledge derives from curiosity about the world. It was scientific curiosity that stimulated us to look for subatomic particles, to investigate the origins of the universe, and to dissect ourselves down to the very tiniest speck of DNA. Since humans and plants continually reproduce, and pass on genetic information from one generation to the next, the messages of the extraterrestrials are very likely to be discovered somewhere within ourselves, and perhaps also in a few species of 'divine plant'. The two conditions I mentioned would then be fulfilled:

- The message would remain indestructible for as long as humans and plants continue to exist.
- The message would only be found by a generation that was capable of investigating molecular biology (genetics) and deciphering genetic codes.

The second premise automatically involves a whole range of other scientific expertise and technological development. No one can study molecular biology without a high-resolution microscope. One has to be able to look into the inside of a cell. No one who knows nothing of the double helix of DNA structure can decipher the genome either. Specific instruments and processes are necessary for all this, which can only be provided by a society which has attained a corresponding level of technological skill. An electron screen-microscope is just as unthinkable without electricity as is overseeing the billions of potential sequences and combinations within DNA without a computer. A whole army of mathematicians would be unable to replace the work of a computer.

These thoughts reveal a further aspect of the palaeo-seti hypothesis which irritates many critics. *Why now?* Why should we suddenly now think of looking for traces of the ETs in human history? To put it bluntly: the universe could not care less *when* we start searching for ETs. But we will start searching for them when we are ready to – which is now. If our science knew nothing about genetics for another hundred years, we would not possibly be able to start looking for genetic traces of extraterrestrials until then.

I have written in many books about the evolution of human beings from the race of hominids.[2] The most recent discoveries of conservative anthropology just make me smile. The newspapers tell us that fossil research has now shown that the 'generally accepted theory' of human origin may need to be revised.[3] This is because Chinese scientists have been investigating a pre-human skull that is 200,000 years older than it ought to be according to previous theories. One has hardly finished digesting this piece of news when American anthropologists announce that they have dated three skulls using the most up-to-date methods, and found them to be 800,000 years older than *Homo erectus* (the upright forerunner of man).[4] There is disagreement about whether human beings originated in Africa or Java. Who knows, perhaps they came from China; or perhaps fossil remains will soon be found in Japan that will ruin all the present theories.

I actually think that anthropological research is not investigating an *intelligent* human species, but ape-descendants and mutations. Does it really matter much whether ape-bones are 1.8 or 3 million years old? I am not in the least interested in discovering when exactly an ape species learned to stand upon its hind legs and stretch its toes out straight. I do not dispute that whole branches of the ape family kept changing throughout the last 20 million years, and that our own ancestors descended from the same race. But all this actually has nothing to do with *Homo sapiens* developing intelligence. It was the *gods* who created the intelligent human being. Of course, they took the raw material of the hominid race for this purpose – where else would they find it? And geneticists will discover the genes which these 'gods' implanted in us; the only question is whether they will be allowed to reveal their results, for this would be proof of the palaeo-seti hypothesis. The starter's pistol for the race for truth has long ago been fired. The geneticists, razor-sharp and on the ball – and not all that 'religious' – will provide the proof we need.

Machines to Make Us Transparent

At the end of February 1987, the scientific magazine *Nature* (Vol 325) announced that Japanese geneticists had developed a super-sequencer: an apparatus that was capable of deciphering a million DNA characters per day. Since then, time has not stood still. The Human Genome Project is in full swing. Whenever governments hold back funding because ideological blinkers limit their view, industry steps in instead. In the USA alone there are roughly 1,300 private and semi-private gene companies. A few miles from Washington, the gene-robots, the super-sequencers, work round the clock. There, in the suburb of Gaithersburg, behind a small front garden, one can find The Institute for Genomic Research (TIGR). Thirty of the sequencer machines stand in a sparkling clean hall. The director of TIGR, Dr Craig Venter, is a far-sighted man: he has given his gene-robots mythological names such as Hercules, Thor, Jupiter and Bacchus. The old gods are coming back to life!

'Every day,' he tells me, 'the TIGR robots are deciphering chain sequences of about 600 genes, and about 500,000 base molecules are being stored.'[5] In no more than ten years, every geneticist should have access to the complete human genome. Then the transparent human being will be a reality.

But TIGR is just one fish in the sea of the Human Genome Project. Various universities round the globe have become involved in decoding genetic material, large pharmaceutical companies likewise. In countries in which a regressive political situation has prevented gene research, the multinationals have moved their research elsewhere. In the field of genetics, the old truth of military technology also applies: 'If we do not do it, someone else will.'

And what are they doing? The human being possesses about 110,000 genes which are dispersed among about 3,000 million DNA segments. At the time of writing, about 10,000 genes have been investigated. We know now what their function is. 10,000 decoded genes compared with the 110,000 in the human genome may seem little; but more and more super-sequencers are at work, and computers are continually storing and comparing their 'gene snippets'; and also, the more genes that become known, the easier it is to home in on the others.

How can the layman understand this process of decoding? What is happening? Genes are tiny segments in the DNA double helix. One can also imagine this double helix as a kind of rope-ladder, or a zip whose fastener consists of nucleic acid chains. Every cell in the human body contains a DNA strand; and just as a rope-ladder has rungs, so too does the DNA: four different kinds of chemical composition. These are called adenine, guanine, cytosine and thymine. Together with a phosphoric acid-sugar base, several 'rungs' of the 'rope-ladder' form the nucleotid sequences. These are, as it were, the letters of the genetic code. But the 'rungs' of the 'rope-ladder' do not just stick to the 'rope' without any connection. The nitrogenous base material adenine is 'keen' to combine with thymine; and the guanine 'feels' itself magnetically drawn to cytosine. (In terms of the 'Lego brick' model, every brick does *not* fit every other.) Now imagine the four basic materials in four different colours, and stretch out the whole 'rope-ladder' to a length of 100 metres. In this model, the DNA is the rope-ladder and the colours are the letters of the genetic code.

What then occurs? Within the cell, the DNA opens its 'fastening' segment by segment, 'rung' by 'rung' and begins to double itself. The nucleotids adhere to the basic material they correspond with – the chemical compositions splashing about in the cell, which we ingest through our food and which our organs take up and reduce to their components. Thus a new DNA strand grows which is absolutely identical to the old one. The cell now divides and the new cell again splits its DNA strand and reproduces itself. So a cell-clump grows, and finally forms a body – and in every single cell is to be found the blueprint of the whole. The human being possesses about 50 billion cells, in every one of which his whole 'programme' is contained.

Every 'letter' of the genetic code is responsible for different growth and functions within the human body. If something starts mutating in some portion of the DNA so that, say, liver cancer is produced, it should be possible to cut out the particular genetic strands responsible and replace them with a health-producing combination of genetic material. But in order to do this, the geneticists must first know exactly which combination is responsible for which functions. This decoding is what the super-sequencers are working at.

And why do we need to know about genetics at all? Are we not meddling in God's work? Can we not just be what we are and leave well alone? Because of environmental factors such as radiation and chemicals which enter the cells through polluted foodstuffs, defects arise in the genetic process; perhaps a cancerous tumour suddenly starts growing, that can attack all the cells. Such defects are passed on to following generations. If we want both to heal the sick person and prevent the defective genes being passed on, we have to know which section of the 'rope-ladder' is producing the wrong kind of 'rungs'; then we can start making repairs to the genetic structure. And this is actually already happening.

Today, hormones are produced by genetic means; there are genetically produced insulin, enzymes, proteins, and all kinds of microbes which are used to neutralize spilt slicks of raw oil or to split up harmful bacteria. All sorts of medicines are already produced genetically, for instance anti-inflammatory, anti-depressive and body-building drugs, and vitamins. The food and washing-powder industries, unbeknown to the consumer, have been using genetic enzymes for a long time already. Teenagers proud of their new jeans with the stone-washed effect are unaware that they have to thank genetically engineered enzymes. The age of the gene supermarket is getting into full swing; and a brand new profession is joining the ranks of older occupations: the gene therapist.

Not of This World

But what will the geneticists make of it when they find more and more genetic information attached to the 'rope-ladder' that could not possibly have derived from our ancestors? It is relatively easy to compare, after all; our relatives, the gorillas, chimpanzees and orang-utans are still around. What will people think when they find out exactly which gene-segment is responsible for human speech, and at the same time discover (by comparison with the gene composition of the ape family) that this segment *suddenly* made its appearance, rather than evolving slowly?

And what will people say when human genetic material

comes to light that *cannot* be of earthly origin, because it does not fit with any known life-form? How will they react when geneticists investigate the mummies of ancient Egypt and discover, beyond any possible doubt, that the oldest pharaohs – the ones with very large skulls, who asserted that they were 'sons of the gods' – contain genetic material that could never have come from the earth, material that is devoid of the 'intermediary stages' of evolutionary theory? And what will they stammer and mumble when the very same genetic patterns are found, on the other side of the world, in the pre-Inca rulers, the 'sons of the sun'? We are on the escalator of knowledge and we cannot jump off now. The apocalypse is going to come long before the end of the world, in the shape of our realization about the origin of human intelligence.

But what is possible for the human genome can also be at work in animals. For quite a number of years now, people have been making a good deal of fuss about dinosaurs.[6] Since the film *Jurassic Park*, we are constantly hearing all sorts of theories and 'proofs' about why they suddenly became extinct.

About 200 million years ago there were a very large number of dinosaur types: 12-metre long, carnivorous monsters which lived in Egypt; others with spikes and armour-plated shells; plesiosaurus with its small head and powerful tail-fins, which was adapted to water; and the 30-metre long, 12-metre high brachiosaurus. As many as 100 species existed, including the flying dinosaurs. Then suddenly, without any warning, the whole lot died out about 64 million years ago. And this happened on all continents at once, as though some infection had broken out that just affected the dinosaurs and nothing else. There are no end of theories which try to explain this sudden extinction.[7] The most recent one suggests that it might have been caused by a meteor colliding with the earth; but why, then, were only the dinosaurs affected, and not all other creatures?

In the film *Jurassic Park*, the stomach contents are extracted from a mosquito preserved in amber for millions of years. Because it had stung a dinosaur shortly before its death, its stomach contains some portions of dinosaur genes. These are transformed – hey presto! – into new, living dinosaurs. Such a thing is possible in fantasy, and even in theory; but one would need more basic material than a few snippets from a mosquito's stomach. To manufacture a dinosaur, about 50,000 genes to every thousand cell-components would be necessary. And these are just not available – unless perhaps in a small bird.

Jurassic Sparrow

The Munich palaeontologist, Dr Peter Wellnhofer, carried out investigations into a fossilized primeval bird, the archaeopteryx. This bird is about 150 million years old, 40 centimetres long, and worth £2 million; there are only seven of them in the world, which puts the price up. Between the bird's teeth, Dr Wellnhofer discovered triangular bone fragments which are typical of a quite different species: the carnivorous dinosaur, allosaurus. This convinced him that all bird species, 'from the sparrow to the condor, are descended from the dinosaurs.'[8]

According to previous theories, birds are descended from the reptiles. I am not in a position to judge between differing opinions on this; but if birds are the descendants of dinosaurs, every sparrow would contain inherited gene material from these ancient creatures.

Perhaps the geneticists will also discover why all the dinosaur species had to disappear from the surface of the earth. Perhaps these monsters were some kind of threat to the earth; perhaps they would eventually have eaten absolutely everything – plants and other animals – and so rendered pre-human evolution impossible. Perhaps someone wanted to prevent an ideal planet like the earth – neither too hot nor too cold – being dominated by huge, stupid creatures which offered no potential for developing intelligence and technology. Perhaps, perhaps ...

What about human consciousness? How did it arise? Seventeen years ago, Dr Julian Jaynes, Professor of Psychology at Princeton University, USA, posed this question and was met with a general shaking of heads from his colleagues.[9] Consciousness? That just developed in the course of evolution. Really? But how did we become aware that we are? Is a clump of cells aware of its own existence? Consciousness has nothing

to do with reflexes, fear responses or tail-wagging; neither is it the sum of all memory processes. Consciousness does not arise through either experience or learning. We can feed an electronic brain as much information as we like, but it will not develop consciousness. Jaynes says:

> Our periods of conscious awareness are really much shorter than we imagine. This is hard to realize, since we are actually not conscious of our moments of unconsciousness. Our consciousness overlays these 'gaps' with its broad net, giving us the illusion of consistency and continuity. One can compare non-consciousness with all the objects in a dark room which are *not* illuminated by the beam of a torch.[10]

What, then, does consciousness consist of? How did it arise? This question, like that about mathematical capability, remains unanswered. Only the human being, among all the creatures of the earth, is endowed with mathematical knowledge. The objection that this is quite logical, since we have to be able to count in order to bargain with each other and exchange goods, puts the cart before the horse. *First* we have to have the capacity, and *then* we can make use of it. Animals, after all, have legs and claws; but no dog has yet thought of counting his sausages on his paws. Mathematical ability is the prerequisite for every science. Without it, nothing can be computed and compared. Dr Max Flindt, who engrossed himself in this question, explained it by means of an example:

> Without higher mathematical abilities, we would be unable to land on another planet. Most ordinary people are unaware of the fact that it is impossible to send a spaceship to the moon or Mars without calling on the highest degree of mathematical precision. The same is true of the Shuttle flights and every satellite. The computations necessary for the precise angle of re-entry of the Shuttle into the earth's atmosphere is a perfect example of this – for on it depends the safety of human lives. If the angle is too steep – by even a fraction of a degree – the spaceship will turn into an inferno; if it is too shallow, the spaceship will bounce off the earth's atmosphere and be catapulted out into space. This has much to do with evolution, since it is a fundamental tenet of evolutionary theory that no capacities develop by themselves without being needed at some

> point. There is no compelling reason, though, why mathematics was necessary for the survival of man's forerunners. Animals of all kinds survive without it (though not, for example, without a sense of smell). In space, on the other hand, survival is impossible without mathematics. And what applies to human space endeavours applies equally to extraterrestrials. If the earth was once visited by aliens, these visitors must have been well versed in mathematics. This is why I consider our capacity for mathematics to be an indication that we are not *only* of earthly origin.[11]

The gods created us in *their* image. And suddenly, without even trying to address such questions, we find the answers in our own genes.

Artificial Intelligence

In the early summer of 1993, an unusual kind of group met together in the Austrian town of Linz; a few hundred computer specialists attended the Ars Electronica conference. This was not the normal kind of computer gathering that takes place continually all over the world; the meeting in Linz was concerned with artificial intelligence. Ulrike Gabriel of the Frankfurt Institute for New Medias demonstrated solar-powered cockroaches. These artificial creatures, directed by light sensors, gathered in groups, 'sniffed' at each other, or carried out sudden backward movements whenever they collided with obstacles. For what purpose? The electronic system in these cockroaches was gathering experience.

Tom Ray demonstrated how this functions, with his computer program Tierra ('Earth'). From hundreds of commands he composed an electronic strand, similar to DNA, that recreates or doubles itself. After 24 hours had elapsed, a kind of screen-biotope had formed. *Der Spiegel* reported it as follows:

> First of all a strand multiplied very fast and spread itself explosively through the electronic memory store. Then the first mutations appeared, which were also capable of multiplying and of combating their forerunners. At last, computer parasites also entered the field, which only transmitted half the commands. These parasites occupied the forerunners' program and used their reproduction code. Now the

> electronic mechanisms developed spectral defence reactions, similar to an immune system, which were capable of blocking off the computer viruses before they destroyed the original program. And just as in life, the parasite population was decimated, and the whole process began all over again; except that now the program had been enriched by its experience with the parasites. The computer had vaccinated itself.[12]

These experiments show that artificial intelligence and life are possible. But what about consciousness? That must be a privilege of those life-forms which are endowed with feeling. And feelings, in turn, are connected with bodily conditions directed by hormones. The hormones, in turn, are activated by our perceptions, in which our sense-organs and personal experience combine. Artificial intelligence, on the other hand, knows nothing of hormones. True, it can compare different pieces of information with lightning speed (experience), and on that basis make correct deductions (learning); but it cannot *feel* – unless, of course, we fit it with a feeling body, in which case we would have nothing other than a life-form.

The brain of a computer, with its high-grade chips, is so sensitive to environmental factors – smoke, dampness, fluctuations in temperature, blows, foreign objects or animals (an ant in the works would cause chaos) – that it must be protected by an outer housing. It is no different for life-forms, whose brain is surrounded by the bones of the skull. By receiving and exchanging information, the computer increases its knowledge, just like life-forms, and can continue to do so for thousands of years.

Let us look at a few historical dates in this connection. Human speech arose roughly 30,000 years ago, as the first means of communication. The oldest rock-paintings or carvings, the first *visual* forms of communication, are about 13,000 years old. The first form of script is only 5,000 years old; and 3,000 years ago the first means of long-distance communication arose, in the form of smoke-signals, fire and mirror-signals. Printing was invented 500 years ago, and in the last century, telegraphic communication developed. Only in the last 100 years have we had moving pictures (films), and in the last 30, computers available for everyone.

A very erudite scientist of the 18th century would have read about 200 books; he would have needed to keep up with only a very few specialist papers to remain abreast of his field. Nowadays, world-wide, more than 300,000 newspapers and magazines are in circulation; in addition there are innumerable radio and TV programmes, not to mention the yearly flood of specialist papers, theses and books. The Library of Congress in Washington is stocked with 100 million documents, and all the other libraries of the world contain another 1,000 million.

It is clear that no one can keep track of this deluge of information. And since the life-expectancy of human beings, as well as the thousands of millions of human brain cells that each of us has, is not enough, we now store human knowledge outside the brain. Future generations will probably need to learn less than us, but instead will have to know how and where the information they need is stored.

It must be the same for extraterrestrial life-forms. Either they have brain cells like us, in which case their storage capacity is limited; or else they are a kind of computerized robot, which can draw at will upon the information they need through a still larger computer. A third possibility would be a synthesis of both of these. Natural beings could be raised, with attention paid to their genetic structure, so that they develop an enormous brain capacity which, however, is only used to a minimal degree. Why? The half-filled software capacity of a computer has storage space for new information. A human brain that only uses 20 per cent of its capacity can be 'filled up' with knowledge, as required, if the gods so wish.

They seem to wish it; and this brings me to the central point of my theme. In my last book,[13] I discussed several UFO sightings and touched upon some accounts of 'abduction'. Let me just briefly recap.

Not Right in the Head?

For over 30 years, according to UFO literature, there have been frequent cases of people who assert with absolute certainty that they have been abducted by aliens, medically examined and interfered with in the genital region – not abused or raped, but investigated, as though they were in a laboratory. Male abduction victims were convinced that sperm samples had been taken from them; females spoke of pregnancy tests, 'sucking out' procedures, and even of artificially created pregnancies. In the latter case, the growing foetus was operatively removed some weeks later.

Naturally no one took these reports seriously: we all know that people can harbour secret sexual dreams, wishes and fantasies. And doctors are familiar with the phenomenon of false pregnancy. It is also quite possible that some women may become pregnant but not want to divulge who the father is, so they use the ET excuse – even though no one believes them. Telling such a story might also make one feel special or chosen, or even that immaculate conception has occurred. In the last three decades I have been content to dismiss all such stories as amusing fabrications, not bothering to ask myself what extraterrestrials might want human genetic material for.

But it is very likely that I was wrong; for what seemed the product of weak minds has recently received some methodical substantiation. In 1987 the American author Budd Hopkins presented the results, supported by several scientists, of many years of research.[14] People he interviewed described – sometimes under hypnosis – how genetic material had been 'siphoned off' from them. There are cases in which the same person was abducted three times over the years: at puberty, as a young man and as a 35-year-old. *If* this is true – and I still reserve judgement – it would mean the person had been 'ringed' by aliens, just as we ring birds, dolphins or bears.

Shortly after Hopkins published his research, other authors reported on similar horror stories.[15] Not only individuals but also whole families had apparently been abducted by 'strange lights'. The victims floated about in brightly lit rooms; the men's genital region was covered in some 'rubber-like substance' and subjected to 'sucking movements'. In other cases, they were sexually stimulated by 'a very pretty woman' and even 'mounted'.

Whenever I broached the subject of 'abductions' with my

acquaintances, they started to laugh. Our intellect is not well disposed towards abductions by ETs – still less towards their experimenting upon us in such ways. The whole thing sounds too far-fetched. People who think there are no such things as aliens are not, of course, going to be convinced by such stories. They *know*, with a sleepwalker's certainty, that UFOs do not and cannot exist. They put up total, insurmountable barriers which no argument can penetrate. And people who do think UFOs might exist find the abduction stories weird, grotesque and crazy. They do not see any reason for ETs to behave like this, even if there are such beings.

But I am afraid that we are going to have to rethink our attitude; and this revising of our ideas has a good deal to do with our brain, with the capacity of our grey matter, with genetic intervention as well as with the return of the 'gods' and their prophets.

Dr Johannes Fiebag, trained as a scientist, has investigated recent abduction cases in Germany, Austria and Switzerland,[16] among them the Berlin woman, Maria Struwe. Fiebag describes her as 'a good-looking woman, intelligent, attentive, critical; not reserved, but keeping her own distance from the events she describes'. Maria Struwe recounts a dream – although she was also simultaneously aware that it was not a dream at all. She lay upon a kind of operating table; to the right and left of her stood small alien beings with large heads and big eyes. At this time she was pregnant with her third child – at least she thought she was. She was familiar with all the signs of pregnancy from her previous experience of childbirth, and she had also consulted a gynaecologist.

Then this dreadful 'dream' occurred. The large-headed aliens removed the embryo from her. She woke up in her own bed bathed in sweat, as though she had suffered an appalling nightmare. Shortly afterwards she visited her doctor, who discovered with astonishment that she was no longer pregnant. All symptoms of pregnancy stopped at once. Two weeks later, Mrs Struwe excreted two 'lumps of flesh'. She assumed that these were the remains of the placenta and flushed them down the toilet.

After some time, the Struwe couple decided to try again for a third child. But since all natural means were unsuccessful – in contrast to previous pregnancies – they decided to proceed with artificial fertilization. 'This was to happen on 22 February 1988. But it caused such inexplicable pain to her that the process was halted.' Yet two weeks later, Mrs Struwe excreted two transparent skins, whose origin was unknown. And then suddenly, as though by divine intervention, she became pregnant once more, on 12 May 1988. On 9 January 1989, she gave birth to her third child, Sebastian.

Dr Fiebag suggests various explanations, among which is the following scenario:

- In the summer of 1986 Mrs Struwe was pregnant.
- In the third month of pregnancy, the ETs extracted the embryo.
- The aliens implanted some kind of skin in her womb which was intended to prevent further pregnancies.
- This is why she failed to become pregnant either by natural or artificial means.
- But these 'barriers' were excreted, and normal pregnancy could then follow.

All these events could be explained away as 'an unusual pregnancy' if it were not for Sebastian. This little boy kept telling his parents of strange dreams, peopled by monsters with big heads and large eyes. He relates that he has seen 'little children in boxes'; also that he has been flying in the air and that the monsters have poured fluids into him. They converse with him 'through his lungs' – which probably means in some kind of internal way. When Dr Fiebag showed the boy some drawings depicting various kinds of ETs, he immediately identified the little ones with the large heads and eyes. Mrs Struwe assured Dr Fiebag that she had never spoken to Sebastian about her 'dream', or about extraterrestrials with large heads and outsized eyes.

So what is going on? The researches of Dr Fiebag in German-speaking countries are paralleled by the investigation of Professor David Jacobs in America. He believes that the sperm extractions and artificial fertilizations are the reason for all the abductions, the aim being to create a half-human, half-alien life-form.[17]

The reported cases are increasing; there are not hundreds but thousands of them. The books referred to in notes 15, 16 and 17 are just the tip of the iceberg. So is it just a passing craze? If so, why suddenly now? Have thousands of people who do not even know each other, who live in different continents, suddenly all been infected by the same madness? Do all these cases have a psychological explanation?

Right in the Head After All?

No they do not, says someone whose views we should respect. Dr John E Mack is a top psychologist, a professor of psychiatry at the most renowned university in America – Harvard. Professor Mack is not only a psychologist and psychiatrist, but also an experienced doctor at the Cambridge Hospital, MA, and winner of the highly sought-after Pulitzer Prize. He is 64 years old and therefore no longer belongs to an impressionable breed of young will-o-the-wisps who follow every fashionable craze. He knows his profession and is quick to see through the tricks, lies or fantasies of his subjects. In the autumn of 1989, he was asked whether he was interested in meeting people who said they had been abducted by aliens. His reaction was that 'they must be mad'. At some point or other, however, he encountered Budd Hopkins, whom I have mentioned, the author of the book *Intruders*. This encounter was to change his life.

In the next few years, Professor Mack met hundreds of people 'from various different regions of the country, who had never had any contact with each other'. And because these people appeared to him to be absolutely sane, reasonable and reliable, he began to develop a professional interest in this phenomenon. At last he undertook a thorough study involving 78 people, subjecting them to all the rigorous tests and procedures of his profession. The results of this research are now available in a hefty volume of some 400 pages. The book is called *Abduction*, and is subtitled *Human Encounters with Aliens*.[18]

Professor Mack's reply to his colleagues, and to all sceptics, could not be more dismissive of their disbelief. Yes, he says, the extraterrestrials exist; the abductees are telling the truth, and embryo extractions, sperm-sample taking and artificial fertilization have all taken place. They are not psychological delusions or wish-fulfilment fantasies. According to this Harvard scholar, we are clearly 'participants in a universe that is swarming with intelligent life-forms, from whom we have cut ourselves off'.

The abductions always occur along the same general lines. Small beings with large, black, vertically positioned eyes and greyish skin are suddenly seen moving about in a bedroom as if they have come through the walls. (Abductions from cars have also been known.) The aliens have small nostrils and tiny mouths with narrow lips. There are often curious lights to be seen outside. The abduction victims feel fear and panic and start imagining all sorts of horrific things. But they are calmed, 'made cold' and physically paralysed. Then begins a spectral flight through the window or the balcony door; and although some victims feel that they are being 'beamed up', they sense the currents of the night air around them. They arrive at a spaceship. Some abductees think that they have passed through walls into the alien spacecraft. Inside it is bright; they are laid upon some kind of operating table and investigated with unrecognizable instruments. Hair and skin samples are removed, fine needles and other objects are inserted into their bodily orifices. Around the operating table stand several of the little grey people; but it always seems that only one fulfils the function of 'chief surgeon', while another takes the role of 'translator'. Very seldom, however, is there any actual spoken exchange – communication takes place by means of telepathy.

This 'treatment' by the abductors can be very unpleasant and is described as repulsive. Physical pain, however, is rarely felt, for the aliens neutralize the pain-centre in the brain. After this 'operation', a dialogue often takes place, during which the abductors try, at least in a fragmentary way, to explain their actions to their victims. Some abductees are shown whole shelves full of small embryos, which float in some kind of fluid. They reach home again in the same way that they left, although mistakes do occur: sometimes the victims wake up again in an unknown place, or find that they and their car have been transported several hundred kilometres.

Weird, one is tempted to say. That *has* to be dream and fantasy. But just think for a moment what a semi-intelligent animal feels like when human beings carry out experiments on it – not dissimilar I imagine.

We easily dismiss these stories – they sound too far-fetched and strange; therefore we enlist the help of logic and reason to disprove them. But logic and reason are, of course, limited tools, which are confined to what we already know. Several generations ago, a supersonic aircraft, a radio transmitter, an X-ray apparatus or a hydrogen bomb which can destroy whole cities at a time would all have seemed illogical and unreasonable. Even 50 years ago, it would have been impossible to explain the atom bomb to a scientist. 'That's impossible,' he would have replied, 'for weapons always release energy and this uncontrolled energy destroys the surrounding area. But this atom bomb, you tell me, only destroys everything of a living, organic nature, leaving tanks and concrete buildings unharmed.'

No, present-day logic and reason are not of much help in understanding the phenomena of abductions.

The 'Ringed' People

Why is it likely that at least some of the cases of abduction are true? The reason is the numbers of people who experienced similar things, despite not knowing each other, and not having read or seen books, videos or films on the subject. Then there is the similarity between the accounts of people in different countries and continents, and the thousands of women who were robbed of their embryos in a ghoulish fashion. There are also unexplained scars on abductees, which have not been made by any human doctor. And finally there are the tiny alien implants which were surgically removed from various abduction victims.

Hang on a minute, what was that? Yes. Professor Mack, on page 42 of the American edition of his book, mentions several tiny metal or fibreglass-like objects which had to be surgically removed: small needle-like implants, in one case situated in a man's penis, in another, in a 24-year-old woman's upper nose region close to the brain. Although these curious implants were subjected to chemical and physical tests, the results are inconclusive since we do not know their intended function. Analysis revealed very unusual compounds or alloys, but nothing which could expose their *raison d'être*. This is perhaps roughly similar to humans marking a bear by inserting a ring into his ear-lobe; the other bears may see the ring and sniff at it, without understanding why it is there.

But we are perhaps a little better placed than bears to make something of all this. If we calm our panic and draw upon our powers of reason, we can at least make a tentative analysis of the situation. The ETs, after all, conversed with some of their victims and gave them some small insight into their unpleasant procedures.

According to some accounts, the ETs stated that our planet was threatened by a catastrophe. Indications of what sort of catastrophe this might be are contradictory and unclear. Other versions say that our human conduct is sliding off the rails. Finally, the ETs have also apparently said that our science is developing according to a misguided 'causal principle' – what we ordinary people would term 'logic'. The model of knowledge imparted to us by scholars and scientists is, according to this version, totally up the creek. (This is hardly surprising if you think of evolutionary theory or the religious sciences!) And because of our false view of knowledge, we are developing an erroneous kind of consciousness – one that is trivial and egocentric, concerned only with ourselves as the centre of the universe.

A Trojan Horse

The bulb-headed aliens with black kiwi eyes have only one remedy for this state of affairs: since the human race is not fit for much, they want to create a hybrid! Our basic genetic structure will survive – but only in a compound with their own. Not a pretty idea.

What those slit-mouthed, grey, rubber-skinned aliens are inflicting on the abductees seems to us criminal. Abduction is a major crime, as is sexual abuse. Human rights are being brutally disregarded, medical interventions are taking place without the subjects' permission, and brainwashing and thought-control techniques are being inflicted on people against their will. The grey ETs do not care a damn about our feelings and laws, they are treating us like lesser animals. They are inserting implants into us, controlling the 'ringed' people, providing no logical nor even provable information about their activities, motives or place of origin. The American author, John White, put it like this:

> The aliens always approach us under cover of dark. They never say exactly why they are abducting us. The whole thing seems suspicious to me, like a Trojan horse; and I have to express my concern about what's going on. If the aliens change their ways: if they come in broad daylight and come clean about their good intentions, then I'll be happy to welcome them into human society. If they don't, I'll continue to regard them as sly, thieving, underworld creatures whose disposition is evil even if they disguise themselves as good. And whether they turn out in the end to be of physical, paraphysical or metaphysical nature, has no bearing on this conclusion.[19]

The aliens, it is true, do not make it easy for us to believe in their good intentions. For 30 years at least, there have been documented accounts of abductions, yet the manner and form of the aliens' investigations into us have never altered. The victims are always treated according to a fixed routine; the sperm tests and embryo extractions proceed in a stereotypical way. No medical research team on earth would need to examine so many thousands of people like this. By the hundredth 'patient', at the most, they would have the information they needed – unless of course they were searching for something specific and different in every single individual.

The human race does not, of course, consist of mass-produced robots – we are all individual and different. None of us has the same memories or feelings as the next person – similar perhaps, but not identical, any more than individual fingerprints are. Every person has his own particular set of experiences: suffers, loves in his own particular way, likes particular kinds of music, reads certain newspapers, likes particular radio programmes.

Is this what the aliens are seeking – our disparity and variety of characteristics? Is that why they need thousands and thousands of individuals, sperm varieties and embryos – in order to form a new *race*? Or are they trying to filter out what seems to them to be the *best* material through an exhaustive amount of comparison? I do not have the answer, any more than other investigators; but that does not alter the fact that the aliens are subjecting us to a criminal kind of procedure. On earth people have to abide by the rules of the country in which they are staying. Do not similar standards apply in the universe?

Even if one takes the point of view that the grey ETs are a degenerate race, superior to us in technology and telepathy but in need of a genetic revitalization, we still should not allow them to do this without our consent. At the end of the day we are also intelligent; we have mastered mathematics, have made great scientific and cultural progress. We are not nobodies; so why should we allow ourselves to be treated like dumb animals? I can understand that the ETs may not want to overwhelm us with their sudden appearance, scaring us like a fox among chickens – what I call the 'shock of the gods'[20] – but enough time has now gone by since the first abductions; it is high time to put an end to these 'fly-by-night' episodes, and to give us some explanation of their activities. It is time for the aliens to overlook our vanity and finer feelings and come out into the open.

Human beings do not take kindly to being left in the dark for decades, and being treated like guinea-pigs. Apart from anything else, our consciousness and awareness have changed. Thirty years ago it would have been unreasonable, not to say crazy, to believe in the existence of aliens. In the meantime, every second American has come to think UFOs are real; and in Brazil, two-thirds of the population has. Even five years ago, 45 per cent of the youth of enlightened France subscribed to a belief in UFOs;[21] even in the anti-UFO country of Germany, in which the 'serious' press either fails to report or mocks every reported UFO case, every fifth person thinks there is such a thing. According to the latest study of the Allensbach Institute for public opinion polls, the percentage is higher than this amongst 16–20-year-olds: a third of them accept the existence of aliens.[22]

Human thinking has not stood still; the moon landings and countless TV science-fiction serials have all helped to expand our consciousness. And the innumerable books which deal with the subject of extraterrestrial life have not just been written for the pulper – half of humanity, at least, has taken notice of such things. The much-vaunted democratic ideals of the free world ought to lead the media to give continual updates on news from the ET front. But this does not happen, which makes me start to understand why the bulb-headed aliens with black kiwi eyes behave in the way they do.

We have probably all had the experience of trying to explain something to someone else or to a group of people and not being heard, being met with lack of interest, being sidetracked by irrelevant arguments, being insulted or perhaps just being cold-shouldered. Further attempts to clarify the matter may continue to have no effect. What do we do in such a situation? We withdraw, assuming further efforts at communication to be pointless. Could it not be the same for the ETs? Are they fed up with trying to converse with us because we are too arrogant to listen?

The abduction cases investigated by Dr Mack revealed something of this nature. The extraterrestrials apparently told the abductees that human beings were not yet ready to communicate with them and accept their existence. If they showed themselves openly, we would react aggressively and consider them as enemies. Our behaviour did not allow them to appear before us; we would be seized by panic. Our consciousness was so riddled with religious and scientific misconceptions that it would not be possible for them to approach us openly. And if they were to approach certain individuals, human society would just dismiss accounts of their existence, even if such reports came from a high-up or highly regarded person.

This is only too true. Just imagine what would happen if the Pope or some prime minister announced that he had been communicating with aliens. He would be whisked away in an instant. The same is true for journalists, editors or top scientists: none of them would be believed. 'Extraterrestials? Here? And he thinks he talked to them? Poor fellow, must have a few screws loose!' That is exactly the kind of reception such reports would have. But for how much longer?

Hybrids of the Future

The ugly-faced alien law-breakers announced an impending catastrophe to their abduction victims. This was the chief reason, they said, for their activities. The good news in all this is that the human race may survive, if only as a hybrid (mixture) between them and us. When exactly will this Doomsday occur? The ETs did not mention any dates – they themselves did not seem to know. Does that not sound familiar? All religions, you may remember, emphasize that no one knows the date of the final reckoning. Perhaps the ETs have access to indicators, similar to those used by geologists to predict earthquakes and volcanic eruptions; such information helps, for example, to predict that the San Andreas fault in California *will* erupt, but not exactly when the eruption will take place.

Is it not possible that the sensors and measuring instruments of the extraterrestrial, slit-nostrilled dwarfs – whose technology is, to us, a book with seven seals – may be registering some approaching cataclysm, whose exact dimensions are unknown? If this were true it would excuse their immoral behaviour, for:

- People are not open to receiving such facts; they are too egocentric.
- It is not known how long there still is before the catastrophe overtakes us, therefore urgent action is called for. Later generations will, in the circumstances, show understanding for the necessity of such unlawful behaviour.

In spite of all their – according to our conceptions – immoral and unlawful behaviour towards us, it has always struck me that the aliens have never mutilated or killed any of their victims. They have always returned them safe and sound to their bedroom or car. Our behaviour towards animals is far less considerate.

The idea has recently emerged that these little beings with bulbous heads are not extraterrestrials at all, but time travellers from our own future. It is true that physicists in recent years have shown that the principle of time travel is not beyond the bounds of possibility, but we still have no idea as to how it could be practically realized.[23] Although it is a fascinating idea, I personally do not think that it explains the phenomenon of these little ETs with outsized almond eyes. Just imagine the following situation.

In the year 3000, the time machine exists. The intelligent inhabitants of the earth are small in stature, with grey skin and huge skulls, and have mastered telepathy. They travel into our time in their time machines and discover that humanity, shortly before the year 2000, is faced with imminent catastrophe. They get busily to work collecting genetic material that they implant in their own kind. If they did not do this, then their race would not exist in the future. No, it does not make sense. If the little grey men are descended from us, there is surely no need to safeguard material that they already have! This time-travel idea is not much help in my opinion.

Falsely Programmed?

Various abduction victims, especially those who were abducted on several occasions, no longer feel themselves to be entirely 'earthly'. In spite of retaining a normal, intact human body, they cannot rid themselves of a feeling that their consciousness has changed. They have the impression that they harbour a latent knowledge that extends beyond the earth and the present epoch. This group of abductees say that they have great difficulty in putting these new feelings into ordinary language. They have suddenly come to possess a knowledge of time and space that fills their whole skulls, as though their previously unused brain capacity had received a sudden input of data. It seems to them that they have entered a soaring cathedral full of millions of frescos and fragments, through whose holy spaces the soft melodies of millennia vibrate. Inexpressible. There are no human words or concepts which can render such feelings and visions comprehensible. Everything seems to co-exist simultaneously: on the one hand it is a real, reasonable and clear vision; on the other, too much – much, much too much – interwoven and intermingled, superimposed and ranged one layer upon another, and at the same time interconnected by lightning-quick channels.

Is this a condition close to madness – the inability to cope with or digest a flood of information? Or are data intentionally being implanted in human grey matter so as to give rise to a cosmic consciousness? Is this cosmic consciousness, this quite other perspective on things, intended to enable those who experience it to show their fellow human beings a *new way forward*? Is this 'expanding reason', as I would like to call it, intended to open people's eyes to other realities? It is fairly common knowledge by now that our world consists of more than we can perceive through our senses alone.

The reader of this book will by now have understood that every cell of his body contains the total information (DNA) necessary to the structuring of his body. At the same time, the DNA also contains innumerable fragments – the so-called 'junk' – which have no apparent purpose. They do not form part of any chain or sequence (in the 'Lego brick' model). It is also generally known that only a small portion of our brain capacity is used. Evolution created something that has, up till now, not been called upon. To these scientifically proven facts can be added what has been passed down to us from the ancient religions.

- The gods created man in their image.
- The survivor of the Flood – whether he was called Noah, Utnapishtim or any other name – was a hybrid between human beings and the 'guardians of the sky' (see the earlier reference to the role of Lamech, page 90).

Our genetic material therefore already contains extraterrestrial portions. The little grey aliens know this. All that they have to do is to awaken the 'junk' by making it compatible with the rest of our DNA chains, so that the half-empty brain is flooded with information. Human beings were never *only* earthly. We developed in earthly ways upon the earth; for generation upon generation we evolved religious, political and scientific self-righteousness, radically suppressing the extraterrestrial aspects of ourselves, and imagining ourselves to be at the centre of the universe. Now, however, the day of reckoning is approaching – the wakening bell of consciousness.

I am not surprised by the reports of many abduction victims, who, without ever having read Erich von Däniken, state that the extraterrestrials were here on several occasions in the dim and distant past, and that they helped human evolution on its way. Twenty years ago already, the astronomer James R Wertz worked out that extraterrestrials could easily have visited our planet at intervals of 7.5 times 105 years; in the last 500 million years, therefore, this would mean about 640 times.[24] Ten years later, Dr Martyn Fogg of London University suggested that all the galaxies were probably already inhabited at the time that our earth came into existence.[25]

SETI Without Europe

Year in, year out, unnoticed by the wider world, international SETI conferences have been taking place, which ever-increasing numbers of people attend. At a recent one, organized by the University of California and sponsored, among others, by NASA, over 70 scientific papers were read. Such themes as the following were examined:

- The Galactic Library: SETI and Scientific Education (Andrew Fraknoi, astronomer, Foothill College)
- The Search for Life on Mars: Taking Stock of What We Know (Michael Klein, Jet Propulsion Laboratory, and Jack Farmer, Ames Research Center of NASA)
- SETI Begins at Home: Can We Define and Measure Intelligence on *This* Planet? (Lori Marino, New York University)

- The Search for Extraterrestrial Technologies in Our Solar System (Michael Papagiannis, Boston University)

Most speakers discussed the possible ways in which technology could be used to detect traces of alien life – such as the kind of radio frequencies which might pick up extraterrestrial signals. There was also, however, criticism of too much amateurism in the field of SETI research; many felt that in order to be taken seriously by the wider public, it should exclude the amateurs.

I beg to differ; as far as I can see this attitude just repeats the old, élitist 'only we know best' attitude that has so often led us up the cul-de-sac of narrow-mindedness – whether in the political arena or in the fields of religion or science. Throughout history, the establishment – of whatever kind – has always tried to hold itself aloof from other, ordinary human beings, to exclude them from access to both true and false knowledge. Religions still perpetuate this practice; and political groups still try to guard their pitiful secrets, although these always leak out in the end. Such attitudes are just ways of trying to assure self-advantage by excluding others. How, after all, are new ideas spread? Through whom do they enter public awareness? From whom do new, revolutionary ideas often come? And finally, who finances almost all of science, from archaeology to astronomy?

Élitism has never yet succeeded in preventing the dissemination of knowledge, but it has considerably slowed down the process. Élitism suppresses public consciousness and nips fresh ideas in the bud. It is public awareness which brings new ideas into common circulation and forms a seed-bed for their propagation. Public life is the antithesis of secrecy and censure. At the same time, however, I am also of course convinced that the specialists must be allowed to work unhampered by public pressures and intervention, unhampered by what may often be the pseudo-knowledge of amateurs. But they should not try to conceal and shroud in secrecy the flowering of their results. 'Even military courts are unable to silence a rumour' (Johann Nestroy, 1801–62).

Just imagine that all mankind possessed telepathic powers, as we believe the extraterrestrials do. In a telepathic society there can be no secrets or élitist knowledge; this has clearly not harmed ET society.

At the last international SETI conference, 73 intelligent lectures were held, but not a single one about UFOs, abductions or even the palaeo-seti hypothesis. Such themes are regarded as unworthy of 'real' scientific research; as if there were not also scientific publications in the UFO field, written by down-to-earth experts, based on proper research (for example, *Present UFO Research*, by the physicist Illobrand von Ludwiger[26]). And what about the Harvard professor Dr Mack? Is he now to be suddenly excluded from the ranks of scientists?

Why do those who devote themselves to the search for extraterrestrial life exclude the most relevant themes and people from their considerations? How can a respectable branch of science – which the SETI group has now become – allow itself to be so prejudiced as to ostracize certain other paths of investigation? Does science not rely upon a broad base of information? Without UFOs and palaeo-seti philosophy, the scientific discipline of SETI is incomplete, and its results – widely publicized in the media – are half-hearted, not to say amateurish. It is science which accuses amateurs of failing to take all relevant aspects of a theme into consideration, of being one-sided, unbalanced and incomplete. But in this case, I regret to say, the tables are turned: you, my dear SETI researchers, are shutting yourselves off in an élitist ivory tower, and failing to take account of the whole picture.

I actually know why UFOs and palaeo-seti philosophy are not allowed to be a subject of debate in international SETI conferences. Some personal observations are in order here. In 1969, when my first book, *Chariots of the Gods?*, made headlines in the American book market, various prominent and less prominent critics reared their heads. That is fine – criticism belongs to both democracy and the pursuit of scientific rigour. But alongside these critics there were also venomous attacks and even whole books written in an attempt to repudiate my ideas, mainly from religious sources or from conservative branches of science such as archaeology and anthropology. To these attacks were added wholesale lies cooked up in the kitchen of misinformation and fed into the digestive system of the media circus. By these means, a negative image of my ideas was propagated and disseminated, gaining widespread currency among journalists and the like. It was the old story; soon it became completely taboo in scientific circles to say anything positive at all about my work. Curiously, however, my ideas started surfacing in all sorts of publications – but without ever acknowledging their source. The scientific establishment allowed itself to be ruled by prejudice, failing to have the courage to set the record straight.

Things have not improved. More than a quarter of a century after the publication of my book, *Recalling the Future*, palaeo-seti philosophy has been elaborated and documented in a further 19 books and a 25-part television series.[27] There is a wealth of evidence provided by very ancient texts and archaeological remains, as well as books by a variety of authors from many different countries – but none of this makes any impression on the SETI researchers. It is not allowed to make an impression on them – it is more important to protect the élite.

Steven Beckwith, director of the Max Planck Institute for astronomy in Heidelberg, is of the opinion that 'there are many planets in our galaxy which may have conditions suited to the development of life.' And the British astronomer, David Hughes, adds: 'In theory at least, there must be sixty thousand million planets in the Milky Way. Four thousand million of these are likely to be similar to our earth – humid and conducive to life.'[28]

The cosmos is seething with life – including life-forms that are similar to the human race. And at least one of these extraterrestrial civilizations visited our planet millennia ago. That is easily proved, so why do the SETI researchers not want to know? And by the way, the difference between scientists and amateurs often consists only in a little phrase: amateurs are people who get paid nothing for doing a lot, while professionals are people who will not do anything for nothing.

The degree to which the SETI scientists have already allowed themselves to be squeezed into a straightjacket is demonstrated by the Declaration of Principles Concerning Activities

Following the Detection of Extraterrestrial Intelligence.[29] This is a piece of legislation which all scientists officially involved in SETI research are required to abide by. It consists of regulations which dictate how one should react if extraterrestrial intelligence is discovered. I would like to share some of these regulations with you so that you may get a better idea of how the discovery of ETs is approached in international circles.

Submitting to Censorship

> We, the institutions and individuals taking part in the search for extraterrestrial intelligence, recognize that this search forms an integral part of space research, and that it should be pursued with peaceful intentions and in the common interest of all humanity. We are inspired in this search by the enormous importance of delivering proof of extraterrestrial life, even though the likelihood of such discovery may be small.
>
> We remind all involved of the agreement which regulates all governmental activities of research into, and utilization of space ... which also applies to all state-funded groups ... (Article XI)
>
> We confirm the following principles which must be followed in the case of disseminating information about the discovery of extraterrestrial intelligence:
>
> 1 Every person and every government or private research institution, or government ministry, which believes it has received a signal *or other form of proof* confirming the existence of extraterrestrial life, should try to test whether the most plausible explanation does, in fact, provide proof of extraterrestrial intelligence, and is not a natural phenomenon of some other kind, *before any public announcement is made*. If no definite proof for the existence of extraterrestrial intelligence can be provided, the discoverer is allowed to publish his findings under the term 'unknown phenomenon'.
> 2 *Before the discoverer makes any public announcement* that proof has been provided of the existence of extraterrestrial intelligence, he must immediately inform all other researchers and research institutions who are party to this declaration ... The parties to this declaration *will make no public announcement about the discovery* until it is certain that the discovery relates to extraterrestrial intelligence. The discoverer should inform the official authority under whose auspices he is working ...

8 *No reply* may be made to an extraterrestrial radio signal or other signs of alien intelligence, before the necessary international consultations have taken place ...
9 ... If reliable evidence is found of extraterrestrial intelligence, an *international committee of scientists and other experts will be called in*, that will act as central focus for further analysis and subsequent observations. *This committee will also supervise the dissemination of information to the public*. The committee should be composed of members of all the above-mentioned international institutions; other members can also be co-opted ... The International Space Travel Academy will serve as official administrative organ for this agreement and declaration ...

What should we make of all this? Scientists naturally avoid sensationalism anyway. Every major discovery is always tested and tested again before it is published. No one wishes to appear foolish in front of colleagues by having to retract a false discovery. It is of course quite sensible if the International Astronomical Union or the SETI commission No 51 – both mentioned in other places in the document – want to be absolutely certain that there is real proof of the existence of aliens *before* the news is relayed to the world. What seems strange, however, is the requirement to inform all sorts of other committees and commissions before the discovery can go public. In plain English, this amounts to censorship; for even when someone is 100 per cent sure that he has produced evidence for the existence of extraterrestrial intelligence, he is still not allowed to publicize the fact. Before this can happen, the powers which have a monopoly over access to information are to have their say in deciding which particular fragments of truth may be released. One has to ask how this censorship process can be reconciled with the freedom of information guaranteed by law in all free countries of the world.

Yet all the passages in this declaration that are concerned with dealing with the public are ultimately just a waste of paper. We – the masses, the people – have long known that ETs exist!

NOTES

1 Däniken, E von, *Der Götter-Schock*, Munich, 1992
2 Ibid
3 'Welcher Kontinent ist die Heimat des modernen Menschen?', in *Welt am Sonntag*, 20 March 1994
4 'Hat der Exodus früher begonnen?', in *Focus*, No 11, 1994
5 Sanides, S and Gottschiling, C, 'Goldader in Erbgut', in *Focus*, No 15, 1994
6 Most people do not know that 'dinosaur' is a term invented by the British zoologist Richard Owen in 1841, on being given some strange reptile-like bones. He took the Greek words *deinos* ('frightful', 'terrifying') and sauros ('lizard') and combined them.
7 Halstead L B, *Die Welt der Dinosaurier*, Hamburg, 1975
8 'Jurrassic Spatz: Vögel stammen von Dinosaurien ab und nicht von Reptilien. Münchner Paläontologe beendet Expertenstreit', in *Focus*, No 3, 1994
9 Jaynes, J, *The Origins of Consciousness in the Breakdown of the Bicameral Mind*, New York, 1978
10 Jaynes, J, Interview in *Psychologie heute*, March, 1978
11 Flindt, M, and Munn, V, 'Is Mathematical Ability Extraterrestrial?', in *Ancient Skies*, Vol 20, No 3, 1993
12 'Sind Radrenner lebendig?', in *Der Speigel*, No 25, 1993
13 Däniken, E von, *Wir alle sind Kinder der Götter*, Munich, 1987
14 Hopkins, B, *Intruders*, Ballantine, 1987
15 Strieber, W, *Communion*, New York, 1987; *Transformation: The Breakthrough*, New York, 1988
16 Fiebag, J, *Kontakt: UFO-Entführungen in Deutschland, Österreich und der Schweiz*, Munich, 1944
17 Jacobs, D, *Secret Lives: Firsthand Documented Accounts of UFO Abductions*, New York, 1992
18 Mack, E, *Abduction: Human Encounters with Aliens*, New York/Toronto, 1994
19 White, W J, 'Aliens Among Us – a UFO Conspiracy Hypothesis in a Religious Mode', in *Mufon UFO Journal*, No 286, February 1992
20 Däniken, E von, *Der Götter-Schock*, Munich, 1992
21 *Science & Vie Junior*, January 1991
22 'Jeder fünfte Deutsche glaubt an UFOs', in *Die Welt*, 28 February 1991
23 Meckelburg, E, *Zeittunnel: Reisen an den Rand der Ewigkeit*, Munich, 1991; *Transwelt: Erfahrungen jenseits von Raum und Zeit*, Munich, 1992

24 Wertz, J R, 'The Human Analogy and the Evolution of Extraterrestrial Civilisations', in *Journal of the British Interplanetary Society*, Vol 29, Nos 7–8
25 Fogg, M J, 'Temporal Aspects of the Interaction among the First Galactic Civilisation. The Interdict Hypothesis', in *Icarus*, Vol 69, 1987
26 Ludwiger, J von, *Der Stand der UFO-Forschung*, Frankfurt, 1992
27 A video of this series, *Auf den Spuren der All-Mächtigen*, can be ordered direct from the Ancient Astronaut Society, CH-3803 Beatenberg, Switzerland.
28 'Planeten-Brut aus dem Urnebel', in *Der Speigel*, No 22, 1993
29 This was accepted in April 1989 by the Board of Trustees of the Academy, and the Board of Directors of the International Institute of Space Law.

5

The Great Deception: Conspiracy of Silence and the Latest Research

'The more you know, the more you doubt.'
(Voltaire, 1694–1778)

My book, *The Eyes of the Sphinx*,[1] appeared four years ago. In it, I examined the unsolved riddles and mysteries of ancient Egypt, and also discussed several theories about the building of the Great Pyramid.

Since then, new discoveries have come to light which I cannot keep quiet about. What is their connection with the theme of this book, with the 'Second Coming' and the return of the extraterrestrials?

The ancient Egyptians regarded Enoch as the builder of the pyramids. (Enoch, Idris and Saurid are the same figure, according to Arabic tradition.) Enoch wrote over 300 books, which he entrusted to his son Methuselah, in the hope that the latter would pass them on to the 'future races of the world'. None of these books has yet been discovered. Have they perhaps been concealed in airtight chambers of the Great Pyramid? May we find there the answers to our questions about the day of judgement and the return of the gods?

And is someone trying to keep this secret hidden from the world?

In the last two years the events surrounding the Cheops Pyramid in Egypt have clearly shown how simple-minded the scientists think people are, and to what extent the media are manipulated, and in turn manipulate public opinion. On 22

March 1993, at exactly 11.05am, a sensation of the highest order occurred. Something took place that was unexpected, unthinkable, and beyond the comprehension of all classical Egyptologists. A bomb could not have made more of an impression on Egyptological perspectives. And yet these shock-waves were channelled, damped down, rendered harmless; and what was very likely a still greater sensation – the event of millennia, comparable with the discovery of extraterrestrial intelligence – was blocked and prevented. So what events were these?

The German engineer, Rudolf Gantenbrink, born on 24 December 1950 in Menden, had accomplished a stroke of genius: a small, technically extremely refined robot of his invention had, after passing 60 metres along a previously unknown pyramid shaft, arrived at a door upon which were mounted two metal clasps. The robot had been rattling along this narrow shaft for the past two weeks, and had continually come up against obstructions that it had to circumvent. Several times it had had to be electrically ordered back to the starting point, so that small technical alterations and refinements could be made to it.

Gantenbrink's robot weighs 6 kilograms and is a caterpillar-type machine, only 37 centimetres long. It is driven by seven independent motors, whose micoprocessors are directed by remote control. At the front are two small halogen headlights, as well as a mini-video camera, of the Sony CCD type, that can swivel and tilt. Although it has a light, aluminium structure, it is capable of carrying a weight of up to 40 kilograms, thanks to the specially made rubber caterpillar-tracks which can find footholds on both floor and ceiling.

Rudolf Gantenbrink has himself been responsible for all the decisive aspects applied in the development of this unique apparatus. He built it himself; the mechanical precision work took months to complete, a good deal of sweat, and £100,000 which he himself invested in this masterpiece of engineering. He received technical support from the Swiss firm Escap in Geneva (specialized motors), from Hilti Ltd in Vaduz (drill technology) and from the firm Gore in Munich (specialized cable). Gantenbrink's robot is a wonderful example of what can be achieved if, instead of saying 'that will never work', one applies a combination of intelligence, technology and will-power.

And what led Rudolf Gantenbrink to think it worthwhile to put all this time and energy into penetrating the Great Pyramid? Surely everyone knew that there was nothing more to find there. The radio and TV reporter Torsten Sasse from Berlin questioned him, and received this reply:

> The whole thing began when I was in Egypt during the Gulf War. I had suggested to Professor Stadelmann (of the DAI, the German Archaeological Institute) that it would be worth taking a closer look at these 'ventilation shafts' – which is what they were still called at the time – since we now possessed technology that would enable us to do this. And because these shafts were also the last part of the pyramid to have remained unexamined.
>
> In 1992 we investigated the upper shafts with a video camera, and set up a ventilation system to see whether the fresh air would emerge at possible outlets. Already in 1992 we ascertained that these shafts definitely emerged somewhere. But we did not know where or how. This was the starting point of all my investigations.
>
> The subsequent project was called Upuaut 2; I must explain this name to you. The robot was so christened at the suggestion of Professor Stadelmann – Upuaut is an ancient Egyptian god and means 'Opener of the Ways'.
>
> Upuaut 2 was developed solely in order to investigate the two lower shafts.[2]

Which 'lower' and 'upper' shafts are we talking about here? The Great Pyramid contains three chambers; and in Professor Rainer Stadelmann's opinion, this is the case in all Egyptian pyramids. Stadelmann is thought of as the 'inventor' of the 'three-chamber theory'. Every tourist who makes the effort to climb into the Cheops Pyramid can visit two of these chambers: the upper one is called the King's Chamber – rather hopefully, since no mummy was ever found there – and the other, somewhat smaller, is named the Queens' Chamber. From the higher chamber, two shafts lead diagonally upwards. These have been called air shafts. In these, Rudolf Gantenbrink set up his ventilation system. Tourists noticed this in the fresh air that flowed back into the King's Chamber – but only for a short time; the system no longer functions. This has nothing to do with Rudolf Gantenbrink, but with the pyramid attendants, who for some reason best known to themselves continually forget to turn it on.

From the lower, smaller chamber, two shafts also lead off: one directly south, the other in a northerly direction. The shaft openings are therefore opposite one another, and at the same height as the end of the entrance tunnel. Rudolf Gantenbrink's robot entered the south shaft. The third chamber lies in the rock beneath the pyramid. It is called the Unfinished Chamber.

What do the specialists think is the reason for the shafts leading from the Queens' Chamber?

They are not agreed. Some thought them to be 'soul shafts', others 'model corridors', and finally they were believed to be the mouths of ventilation shafts,[3] or air shafts. This last idea was nonsense, however, since the shafts were only opened up in the last century, by breaking open the walls. In 1872, the Englishman W Dixon was trying to locate hidden chambers by knocking at various parts of the chamber walls and listening to the depth of tone. When he found a hollower sound, he grabbed his pick-axe and revealed the openings of the 'air shafts' a few centimetres beneath the stone surface. Both shafts have square proportions, 20 x 20 centimetres.

Two things, at least, are crystal clear: first, they cannot be air shafts, for these would have had to lead into the chamber to function; and secondly, they must have been part of the original plan of the pyramid; it would have been impossible to carve them in or hollow them out after the pyramid had been built. Not even a child can fit through a gap 20 centimetres square.

The two shafts of the Queens' Chamber do not lead diagonally upwards like those of the King's Chamber. They first enter the wall horizontally, then begin to climb at an angle of exactly 39 degrees, 36 minutes and 28 seconds. Most Egyptologists were agreed that the shafts 'end after a short distance' – until Rudolf Gantenbrink's robot Upuaut suddenly proved them wrong.

The Opener of the Ways

On 22 March 1993, it was as hot as usual on the pyramid plateau of Giza; and inside the Great Pyramid it was as humid as ever. Rudolf Gantenbrink had set up an improvised table in the Queens' Chamber, made from two boxes and boards. Upon this stood his electronic 'station' and a monitor that relayed crystal-clear pictures from the robot's camera. A video machine was also set up to record the film sequences. While a colleague carefully fed the specialized, very thin and light-weight cable into the shaft, and an Egyptologist from the Egyptian Ministry for Ancient Monuments watched the screen with ever-increasing astonishment, Gantenbrink directed the robot's small steering lever with total concentration. The whole team was under pressure of time, since the Ministry for Ancient Monuments had decided to bring a halt to these investigations that very day. Too many travel agents were complaining because they were not able to take tourists into the Great Pyramid while this was going on. The ministry was also losing money, since entry to the pyramid is not free.

Metre by metre, Gantenbrink's miniature monster crept up the steep passage. The headlights on the front illuminated scenes that no one had seen for at least 4,500 years. Cheops, who was said to have built the pyramid, ruled from 2551 to 2528 BC.

The slow journey passed smoothly polished walls; the robot had to surmount small piles of sand and manoeuvre itself cleverly over fragments which had fallen down from the 'ceiling'. At last, after 60 metres, came the first surprise: on the floor lay a broken-off piece of metal. Shortly afterwards, the great sensation. The robot camera relayed a kind of door or partition, which closed off the whole shaft; on the upper part of the door were two small metal clasps, the left of which was partly broken.

Rudolf Gantenbrink steered the robot towards the door, aiming the laser beam at its lower edge. The red beam, 5 millimetres across, disappeared under the edge of the door. This showed that there was space beyond it. At the lower right-hand corner of the door, a portion of stone was missing. The robot camera relayed dark dust there, which had no doubt blown out of this tiny opening in the course of thousands of years. But the robot's journey had now come to a halt.

Michael Haase, a mathematician from Berlin, worked out the position of the mysterious door.[4] It lies in the south side of the pyramid, at a height of about 59 metres above ground, between the 74th and 75th level of stones. If the shaft which is blocked by the door continued at the same angle, it would reach the outer wall of the pyramid at a height of 68 metres. The horizontal distance from the door to the outer wall is about 18 metres. Rudolf Gantenbrink naturally climbed up the south wall to investigate, but there was no sign at all of any exit to the shaft.

The Sensational News Concealed

The discovery of a 60-metre long shaft inside the pyramid is one sensation, the door blocking it off another. One might have thought that Gantenbrink's efforts and achievement would be properly valued by Egyptologists as a once-in-a-century discovery. If an astronomer discovers a new star or comet, it is common for it to be given his name. That is why I now call the 'new' shaft the Gantenbrink Shaft, as do my colleagues. The narrow-mindedness and jealousy of the Egyptologists, on the other hand, takes a different view. Others, they say, had already suspected that the shaft existed. That is only a quarter of the truth: it is true that people knew of the existence of the horizontal shaft-openings which lead from the Queens' Chamber in both northerly and southerly directions, but no one was aware of a 60-metre passage inside the pyramid. On the contrary, people blustered about 'soul-shafts' which 'end after a short distance'.[5] And theories are not discoveries. All sorts of things may be suspected. But it is only the German engineer Rudolf Gantenbrink who discovered the 60-metre passage and the door at the end of it.

Gantenbrink himself is not interested in sensationalism. His prime concern is to preserve ancient remains. At the same time he wants to inject archaeology with new life, and rejuvenate it by means of new technology. He is a hard-working, honest fellow, who likes to solve puzzles, and who places his experience and brilliance at the service of a fascinating science. But none of this, apparently, is appreciated: Gantenbrink was given the cold shoulder.

After the discovery of the Gantenbrink Shaft, nothing at all happened for a while. Although the specialists in Cairo and in the German Archaeological Institute (DAI) knew of the discovery, there was nothing but icy silence from them. The public were not informed. No one was allowed to say anything. And the public would still be in the dark to this day, if chance and Gantenbrink himself had not done something about it. Gantenbrink showed a copy of the extraordinary video that the robot had taken to a few colleagues; then the British press got wind of it and, only two weeks after the discovery(!), published a tiny article entitled 'Portcullis Blocks Robot in Pyramid'.[6] This article also reached Cairo via fax.

What was the reaction? The DAI in Cairo denied the news. 'That is absolute rubbish,' said the Institute's press officer Christel Egorov to the news agency Reuter.[7] According to her, the discovered passage was only an air shaft, and the mini-robot had just been measuring humidity. It was, she said, well known that there were no other chambers within the pyramid.

Not only may we *feel* deceived by this, but we *are* being deceived! The archaeologists of the DAI in Cairo were fully aware that their statements were false. The robot that journeyed down the Gantenbrink Shaft had absolutely no instruments on board for measuring humidity.

It gets worse. Dr Rainer Stadelmann, the great chief of German Egyptology and director of the DAI, denied that there was any possibility of a secret chamber behind the shaft door. He said to journalists: 'It is common knowledge that all treasures concealed within the pyramid were plundered long ago.'[8] His colleague, the Egyptologist Dr Günter Dreyer, backed him up: 'There is nothing behind that door. It's pure imagination.'[9]

Before I relate how the esteemed circle of Cairo Egyptologists got rid of Rudolf Gantenbrink, I must elaborate on the opinions people have about the interior structure of the pyramid.

It is nonsense to assert that there is nothing inside the pyramid apart from the three known chambers, and that there could not possibly be anything behind the door. If the archaeologists of the DAI were to say that *it is not known* whether there is anything behind the mysterious door, they would be right. But to claim categorically that they know there is nothing there is not only dogmatic and unscientific, but also – to echo the words of the DAI itself – 'absolute rubbish'.

Knowledge of the Ancients

Let us backtrack a bit and have another look at history. In the 14th century, there were ancient Arabic and Coptic fragments lying in Cairo libraries, which the geographer and historian al-Makrizi compiled in his work *Hitat*. There we can read:

> Whereupon the builder of the pyramids had 30 treasure chambers of coloured granite fashioned in the western pyramid: these were filled with instruments and images made from precious stones; with tools of the finest iron, such as weapons which do not rust; with glass that can be folded without breaking; with strange talismans; with all kinds of simple and compound medicaments; and with deadly poisons. In the eastern pyramid he had the various spheres of heaven and the planets depicted, and images made of his forefathers' creations; also incense that is offered to the stars; and books about these. One also may find there the fixed stars and what comes about in their progression from one epoch to another ...
>
> And into the coloured pyramid, finally, he had the corpses of the truth-tellers and soothsayers brought, in coffins of black granite; and beside each soothsayer lay a book, in which were written all his wonderful arts, his life's story and the works he had accomplished.[10]

And who is supposed to have had these mighty buildings erected? Cheops, as the Egyptologists assert? The *Hitat*, as I have already mentioned, tells us:

> The first Hermes, called the Threefold One because of his qualities of prophet, king and wise man (he whom the Hebrews call Enoch, the son of Jare, the son of Mahalelel, the son of Kena, the son of Eno, the son of Seth, the son of Adam – whose name be blessed; and

> whose name is also Idris), read in the stars that the flood would come. Then did he have the pyramids built; and had hidden in them treasures, learned texts and everything which he feared might otherwise be lost, so as to protect them and preserve them.

It is not only in the *Hitat* that Enoch is named as builder of the great pyramids. In the 14th century, the Arabian traveller and writer Ibn Battuta says the same thing:

> Enoch erected the pyramids before the flood, to preserve within them books of knowledge and science, as well as other valuable objects.[11]

It is hardly necessary to mention that Egyptologists dismiss these Arabic traditions. They are sure that the builder of the pyramids was Cheops, even though there are so many convincing arguments against that point of view. I have gone into this in detail in my book *The Eyes of the Sphinx*.[12]

The archaeologists behave as though they were deaf, blind and dumb. I can just about accept, though reluctantly, that they may not want to heed 14th-century texts. But the fact that they also reject the evidence of modern science, if it does not accord with their holy doctrine, seems to me unbelievable. Examples from the last 25 years speak for themselves.

In 1968–9 the winner of the Nobel Prize for physics, Dr Luis Alvarez, undertook an examination of the Chephren Pyramid using radiation. Alvarez and his team made use of the well-known fact of physics that cosmic radiation is constantly bombarding our planet and, as it penetrates solid matter, loses a proportion of its energy. Exact measurement can reveal how fast the protons are penetrating a layer of stone. If the stone contains hollow spaces, the protons are not hindered in their passage to the same extent. Alvarez measured the paths of over 2½ million particles with the aid of a transmitter and an IBM computer. But the oscillographs showed a chaotic pattern, as if the particles were curving around the earth. It was baffling and infuriating. The very expensive experiment, in which several American institutes, IBM and Cairo's Ain-Shams University were involved, ended without any clear results. The head of archaeological research at that time, Dr Amr Gohed, told journalists that the findings were 'scientifically impossible'; he added that either the 'structure of the pyramids is chaotic' or there is 'some mystery here that we have not explained'.[13]

The archaeologists generally ignored these baffling results.

Dating the Sphinx

In 1986 another attempt was made with new instruments and methods to search for hidden chambers in the Cheops Pyramid. The two French architects, Jean-Patrice Dormion and Gilles Goidin discovered various hollow spaces inside the pyramid with the help of electronic detectors. But this did not alter the entrenched opinion of the Egyptologists. Since one of the sponsors of this investigation was the French electricity board, the research was written off as an advertising stunt.

The next major investigation was by a Japanese team of scientists from Waseda University in Tokyo. Using the most advanced electronic equipment, the Japanese specialists X-rayed both the interior of the Great Pyramid and the whole surrouning area including the Sphinx. They found clear indications of a whole labyrinth of passages and chambers inside the Cheops Pyramid. They presented their results in a report that was a model of scientific procedure.[14] And what did the Egyptologists say? That this research was, of course, just publicity and promotion for the Japanese electronics industry!

The DAI team in Cairo are apparently not interested in anything at all. And their colleagues in Europe and elsewhere generally know almost nothing about what happens on the Giza plateau. If it was up to Egyptologists, no research would need to be done, for they know everything already!

In 1992, the geologist Dr Robert M Schoch of Boston University's College of Basic Studies, together with other scientists, undertook geological measurements and analysis of the Sphinx. The results showed that it is at least 5,000 years older than was previously thought.[15] People generally believe that the pharaoh Chephren (2520–2494 BC) built the Sphinx. This is not because any real proof has been provided, but because the name 'Chephren' is still just about decipherable on a crumbling cartouche, if one is determined to read it in that way. This half-erased name does not even belong to the Sphinx but to a stele of the pharaoh Thutmosis IV, who ruled more than 1,000 years *after* Chephren, from 1401 to 1391 BC.

But how did Schoch arrive at his opinion that the Sphinx was at least 5,000 years older than Chephren? His team planted a number of seismic receptors in the ground. Then sound-waves were generated, which allowed a survey to be made of what was beneath the surface, a method which has continued to prove useful in the field of geology. Computers worked through the data and spewed out long strips of drawings, which reproduced an exact underground plan of the Sphinx. There were very clear traces of weathering at a depth of 2.4 metres, which were absent from the rear end. But at that rear end, repairs had been carried out long after the Sphinx was built. During his rule, Pharaoh Thutmosis IV had had the Sphinx dug out of the sand and repaired.

The geological measurements and chemical analysis pointed to only one possible conclusion: the strong signs of erosion and weathering stemmed from a time of prolonged rainfall, which had not occurred in Chephren's time. Like tree-rings, it was possible to date this erosion to at least 7000 BC.

And the archaeologists' reaction to Schoch's data? A storm of indignation. At a conference in Boston, Mark Lehner of Chicago University described Schoch as a 'pseudo-scientist'. Lehner's main argument was as follows. If the Sphinx was really so old, there must have been a culture at that time that was capable of erecting such a work of art. But in those days human beings were just hunters and gatherers. Finito!

It is human nature perhaps, when one runs out of reasonable arguments and one's back is up against the wall, to use insult and abuse. This is, at any rate, what happened in the debate between the archaeologist Mark Lehner and the geologist Dr Robert Schoch. Lehner accused his scientific colleague of 'suspect credibility'. Why this unfair attack? One of the sponsors of Schoch's geological investigation was a certain John Anthony West. And Mr West was guilty of two heinous crimes: first he was not a scientist, and secondly he had already published books in which he posited the existence of a civilization 'older than any we so far know' – sacrilege, in the eyes of a 'real' archaeologist.

The archaeologists are not interested in the fact that Schoch was by no means the only geologist who was involved in the seismic measurements on the Giza plateau. Among the team's members were also Dr Thomas L Dobecki, two other geologists, an architect and an oceanographer. No one paid any attention to their firm conviction that the lowest parts of the Sphinx clearly contained water channels which could only have been formed as a result of long exposure to water. Dr Schoch's geological analysis was roundly condemned by the current director of antiquities at Giza, the Egyptian Dr Zahi Hawass, as 'American hallucinations'. According to him, there was 'absolutely no scientific justification' for Schoch's new dating of the Sphinx.[16]

So it seems that Egyptologists have no interest in results which do not suit them, even if they are scientific ones that have been obtained by proper scientific methods. *They* determine what the world should believe. They do not notice that they are actually sawing the branch on which they are sitting. Public opinion is tired of trusting in science; and a branch of science which only accepts other branches when they confirm its own views deserves little trust.

Another of the exact sciences is physics; and at the Swiss Technical College (ETH) in Zurich, Professor Dr W Wölfli is recognized as an authority. He has perfected the long-disputed process of carbon dating, by means of which the age of organic material can be measured. Professor Wölfli, together with several colleagues from other universities, analysed 16 different materials from the Cheops Pyramid, among which were charcoal remains, splinters of wood, straw and grass fragments. The result? All the samples were an average of 380 years older than the Egyptologists had deduced from the list of kings. One sample from the Cheops Pyramid was in fact 843 years older than it was thought to be.[17]

The physicists had examined a total of 64 organic samples, and applied a variety of methods. *All samples without exception* produced datings that were several centuries older than the ones favoured by Egyptologists. But no conclusions were drawn, no new perspectives were considered. On the contrary: the old position was if anything cemented with new excuses. And if you think 'excuses' is a harsh judgement, I personally consider it too mild a term for the rubbish that we are expected to swallow whole.

Discrediting Gantenbrink

The Egyptologists of the DAI want Rudolf Gantenbrink off their hands. Why? Did he not make a spectacular discovery with his robot? Did he not spend much time and money in the service of archaeology, helping to advance the state of knowledge? Was he unscientific? Not at all – his results can be repeated by anyone at any time. Was he impolite or unfriendly? By no means. Gantenbrink is a very pleasant kind of person. Did he start all sorts of unscientific rumours and speculations? No again; he spoke to the media in a very cautious, reserved fashion. He always stated clearly that nobody knew whether there was anything to be found behind the stone doorway in the newly discovered shaft; he refused to speculate about it. So what did he do wrong? Why is he *persona non grata* to the Egyptologists of the DAI?

He spoke to the press. He did not run to journalists and trumpet his discoveries to the world; the journalists got wind of his phenomenal discovery through British scientists, and sought him out. It is, after all, the job of journalists to follow up and investigate interesting trails. But Rudolf Gantenbrink did not try to get any mileage out of this fact – he remained measured, factual and decently circumspect. Should he have lied and led the journalists up the garden path? Gantenbrink is not a politician!

In a report from the German Press Agency (DPA) of 27 June 1994, the journalist Jörg Fischer writes:

> Once more, as on many occasions in the past centuries, the gigantic pyramids of Giza are at the centre of mysterious and mystical imaginings ... The robot expert Rudolf Gantenbrink from Munich

> independently announced his discovery to the press and said that he suspected there was a burial chamber behind the doorway. 'Some German tabloid has already found the ashes of a pharaoh and a treasure of gold', commented DAI director Professor Rainer Stadelmann, about the 'nonsense' he says has been written on this subject.[18]

The words here attributed to Gantenbrink are unfounded. He never expressed the view that there was a burial chamber to be found behind the doorway. The media, who know no better, have here been roped into the service of a professor who wants to discredit and sideline Gantenbrink's work. Gantenbrink never 'independently' volunteered information to the press, for he was never a member of the DAI and therefore never subject to any information restrictions that this body might have tried to enforce. The DPA report that was passed on internationally and formed the basis of many newspapers' reports achieved the professor's aim of misinformation. People were supposed to believe that Gantenbrink was publishing unscientific imaginings. This in turn so annoyed the Egyptian government that it withdrew permission for further research into the pyramid shafts.

Scholarly Error

This all becomes clearer later in the DPA report:

> The archaeologist [Dr Rainer Stadelmann] categorically excludes any possibility of a chamber: after examining the pictures relayed by a remote-controlled video camera and comparing them with what is known of three other pyramid shafts, he believes his opinion to be confirmed that the shaft is a 'model corridor'. The opening which leads upwards from the Queens' Chamber was intended, in accordance with the religious beliefs of ancient Egypt, to allow the soul of the pharaoh to ascend to heaven. The black dust lying in front of a stone block at the end of the passage comes, according to Stadelmann, from the eroded metal clasps on the 'model doorway'.
>
> His sobering theory and the fact – which he has constantly repeated – that people would be quite unable to crawl through the narrow shaft, let alone hide a sarcophagus or a treasure there, has been largely ignored.

Anyone who does not share the professor's theory is of course living in cloud-cuckoo land. I can understand why he 'categorically excludes' the possibility of another chamber. It was he, after all, who came up with the 'three-chamber theory'. The discovery of another chamber would not fit well with this. One realizes what lengths he has gone to to safeguard his theory if one considers that the empty spaces we know of in the pyramid amount to 2,000 cubic metres, and that the Great Gallery takes up 1,800 cubic metres while the other chambers share the remaining volume between them, but that the Great Gallery is not allowed to be considered a 'chamber'.

And what about the 'model corridor' idea? Let us just think about it for a moment. The ancient Egyptians built the most perfect structure in the history of the world. It consists of about $2^1/_2$ million stone blocks. The initial planning must have been phenomenal: all the blocks and struts fit perfectly and exactly together; it is a building for eternity. Within the pyramid a passage is placed, which we now call the Great Gallery. It leads diagonally upwards to the King's Chamber, is 46.61 metres long, 2.09 metres wide and 8.53 metres high. Since the passage walls incline towards each other as they rise, the roof – of horizontal stone slabs – only measures 1.04 metres. The gigantic granite blocks on either side of the 8.5-metre vault do not lie in the horizontal plane, but, as if to shake us out of our complacency, follow the rising angle of the Great Gallery. The workmanship of the blocks and slabs is of such perfection that the visitor is hard put to find any cracks or joins. Before one reaches this Great Gallery, one has to bend one's back and creep through the ascending passage.

We still do not know why the builders first made a narrow, low passageway that led into the Great Gallery. But Professor Stadelmann, with a somnambulist's certainty, knows that the Gantenbrink Shaft is a 'model corridor', from 'comparison with three other shafts in the pyramid'. Holy Osiris! Where in the Great Pyramid are there any other 'model corridors' that one could compare it with? They have always been known as 'air shafts'!

The Gantenbrink Shaft is supposed to be far too small to allow a sarcophagus, let alone a treasure, to be transported through it. But why then is there a granite sarcophagus in the King's Chamber with dimensions greater than the ascending passage? According to Professor Stadelmann's logic, it ought not to be there.

In this miracle of building, intended to survive to the end of time, the architects of ancient Egypt were supposed to have introduced a 'model corridor'. Yet it is concealed, and actually does not lead directly out of the Queens' Chamber. The connecting openings were only knocked through by Mr W Dixon about 120 years ago. Through this 'model corridor', the soul of the pharaoh is meant to fly up to the stars. The only problem with this is that a pharaoh never lay in the small Queens' Chamber. And even if a corpse had been deposited there and the shafts had been open from the beginning, the pharaoh's soul would not have had clear access to the firmament. According to the Egyptologists, the Gantenbrink Shaft is blocked off by a stone behind which there is nothing. Poor pharaoh!

The 'sobering' theories of the renowned Egyptologists, and the repeated statements that people could not have crawled through the narrow shaft, let alone have hidden a sarcophagus or treasure at the end of it, edge very close to being complete nonsense. Let us consider another possibility, another way of looking at the whole thing. The clever archaeologists only consider that the Gantenbrink Shaft leads *out* of the Queens' Chamber. But what is to stop us thinking that it also leads down *into* the Queens' Chamber from above? Behind the mysterious door of the Gantenbrink Shaft there *may* (not *must*) be a chamber, which has *another*, upper entrance shaft, whose opening may also be walled over just as the shaft opening in the Queens' Chamber was, before Mr Dixon took his pick-axe and broke through into it.

To put it another way, if a robot had travelled *down* the Gantenbrink Shaft from above, it would have come to a halt in front of the wall blocking off the Queens' Chamber, if this had not yet been broken through by Mr Dixon. And all the top archaeological brass would have united in the opinion that there could be nothing more behind it. And no one would have bothered to pierce the apparently final blockage, or to dissolve it with acid. Is that scientific? Where is the curiosity, the striving for knowledge? How can it be stated *a priori* and categorically that there is nothing more to be found behind the door of the Gantenbrink Shaft? And how can everyone who begs to differ be dismissed as a deluded lunatic?

At the end of the shaft, Gantenbrink's robot filmed two metal clasps on the stone door. The fact that they are metal cannot be disputed for, thank goodness, a piece of metal is broken off and lies upon the ground. Since only copper at best was available in Cheops' time, these clasps are called 'copper' with enormous assurance. But they may well not be. Yet Professor Stadelmann and his top-brass Egyptologists have provided a 'natural and reasonable' explanation, which he described to the radio and TV journalist Torsten Sasse:

> What is this [copper handle] for? At first we thought that it might be there for some kind of technical reason. But given its thinness, I would now exclude that possibility, and assume that it is a decorative hieroglyphic sign. And if that is the case, then it has some symbolic content. We must therefore ask what its significance is. It could perhaps be the lotus flower sign, which is symbol of the south. Or, perhaps more likely, the sign *shuut* in ancient Egyptian – that is a kind of sunshade that is carried behind the king when the royal train processes. If this is what they are, then they could be there ready for the soul of the king to make use of when he flies up to heaven.[19]

Good heavens! What a load of unjustified interpolation. The Great Pyramid is wholly anonymous: we know nothing whatsoever about the team of architects and building engineers, or about the priest or pharaoh involved in its construction. There is not a single inscription to give us a clue about how it was built. Nobody left any hint which could help answer a single question about the building of the pyramid. In the pyramid itself there are no hieroglyphs, no walls covered in writing, such as we find in other ancient Egyptian grave-sites. Cheops, the supposed force behind its construction, is thought to have been a despot who had the idea of leaving behind him the greatest building of all times. Yet he and his servants forgot to praise him in either text or image. Not one little inscription was placed there in honour of the pharaoh Cheops, nowhere can one discover any record of a heroic deed of this supposed egomaniac. All the walls, corridors and chambers of the Cheops Pyramid are polished smooth – they have never, now or in the past, been decorated with a single word. Perfect anonymity.

And yet we are supposed to believe that at the end of the Gantenbrink Shaft is to be found the hieroglyph *shuut*, placed there so that the pharaoh can rise to his ancestors without suffering from sunburn? It is an idea that does not – to put it very mildly – strike me as likely!

At the lower edge of the door at the end of the Gantenbrink Shaft, a small triangular portion is missing. It is there that the robot-eye caught sight of a small streak of black dust. Professor Stadelmann thinks this is dust from the eroded metal of the metal clasps.

But let us ponder this for a moment: the learned Egyptologists think the Gantenbrink Shaft is just a 'model corridor' that has been blocked off at the end with a stone; but in this case there would not be the slightest breath of air or wind. It is only the *left* piece of metal that has broken off; yet the dust lies in the *right* corner. Have dust-spirits been at work? And if the metal clasps had quietly rusted over millennia, the black dust should lie along the lower edge of the door, directly beneath them. But it does not. It eddies out of the small triangular hole, as though a very faint draught of air had blown it through. Such a faint draught suggests that the Gantenbrink Shaft extends beyond the door. Or that there is a chamber behind the door into which another shaft leads. The 5-millimetre wide laser-beam of the Upuaut robot also passed *under* the door. Whether it is a door or final blocking stone, it does not lie flat on the shaft floor. Should that not make us think? Obviously not: the Egyptologists have agreed with each other that this is a 'model corridor', so no further research is needed!

Squandering Trust

On 5 August 1993, the director of the Egyptian Museum in

Berlin, Dr Dietrich Wildung, wrote in the *Frankfurter Allgemeine Zeitung*:

> The Egyptologists no doubt have reason to thank the technical expert [Rudolf Gantenbrink]. Yet the latter is unable to resist the temptation to win himself a sensational kind of publicity, and has begun splashing about cluelessly in the swamp of pyramid mysticism and imagined treasure. And lo and behold, here comes Erich von Däniken onto the scene, interpreting the black dust at the lower edge of the stone slab as a sign of the concealed mummy of King Cheops. And where an untouched mummy is to be found, the priceless treasure cannot be far behind, which since Herodotus has stirred the imagination of the world. The automatic mechanisms of trivialized archaeology start up their routine; and the more cautious and careful specialists are dismissed as yesterday's men, who are loth to jettison the ballast of traditionalism and intellectualism.[20]

This is the kind of rubbish with which Egyptologists spin their comfortable cocoon and disparage those who think differently. I never dreamed of thinking that the black dust indicated that the mummy of King Cheops lay behind the slab. This idea came from David Keys, the archaeological correspondent of *The Independent*.[21] I would never have come up with such an idea, since I do not believe that the Cheops Pyramid belonged to Cheops, let alone that it contained a grave.

What, then, do I believe is to be found behind the blockage in the Gantenbrink Shaft? Probably the same as is hidden in all other still undiscovered chambers: texts and documents of all kinds, as was suggested by the Arab historians of the 14th century mentioned above.

David Keys drew people's attention to another curious thing: the vertical distance between the Queens' Chamber and the King's Chamber is 21.5 metres, which is exactly the distance between the Queens' Chamber and the door at the end of the Gantenbrink Shaft. Is this chance, or clear evidence of another chamber?

The DAI experts would now like to investigate the north shaft leading out of the Queens' Chamber. Rudolf Gantenbrink had also thought of that. I personally think one should first finish the task in hand. There have been various suggestions for opening, breaking through or even corroding the door. Why are the opinions, work and knowledge of someone like Rudolf Gantenbrink suddenly not wanted? How can academics, who are otherwise usually very reasonable and open-minded, even humorous, suddenly react in such an eccentric and disagreeable way?

I can only imagine that they are envious. Top archaeologists are hurt in the depths of their psyche because a non-archaeologist has succeeded in making an unexpected discovery. They are bitter because Gantenbrink spoke to the press. Or do they want to conceal what might be found behind the door? Do they want to keep any discovery to themselves, safe from the rabble, and take their time in secretly evaluating it?

What is indisputable is that the scientists in Egypt have no wish to have any public interest or involvement in what they are up to. Any information they dispense is censored by them. They no doubt want no journalist or neutral observer present when the mysterious door is forced open. They want no TV camera relaying pictures of whatever discovery is made to the outside world. They want no one else, from any other branch of science, to analyse the metal clasps on the door. And this childish secretiveness is, the Egyptologists say, only to allow them to pursue their investigations in peace. I can understand this desire; but this is not some insignificant grave. It is the Great Pyramid, which has fascinated humanity for thousands of years. It is the most gigantic building on this planet, a wonder of the world, a monument around which legends and stories have grown up through the millennia. Egyptology is missing its one chance to demonstrate to the world at large that its procedures are correct and scientifically rigorous. It is squandering the possibility of showing the loonies and mystics – who believe there are secrets and conspiracies behind every corner – the unadorned facts; of showing what is actually there, once and for all.

Or are they actually terrified of what they might find at the end of the Gantenbrink Shaft? Archaeologists were not so jumpy in the old days. When the graves of Tutankhamun and Sekhemkhet were opened, journalists were allowed to be present. Since then, global media networks have been developed, which would allow the live images from Gantenbrink's robot to be relayed simultaneously to millions of homes around the world. There would be no need for a horde of journalists to squeeze into the Queens' Chamber and upset anybody's peace and quiet. But *live* pictures are what is needed, shot as the discoveries are being made, not edited pictures which are released days, weeks or months later and provided with an unctuous legend of some sort to make it palatable to the status quo.

Just imagine if the Americans undertook the moon landing in secret, and weeks went by before NASA released censored images to the world. The outcry would be completely justified: 'What are you keeping from us? What have you got to hide? Why should we taxpayers finance an organization that treats us like children?'

The Egyptologists of the DAI and the Egyptian Ministry for Ancient Monuments behave as though openness is a threat. Those who avoid public scrutiny and cloak themselves in secrecy have something to hide. If one starts off by concealing something, one ends up having to perpetuate the deception. As long as the 'information politics' of the Egyptologists expends itself in secretiveness and avoidance tactics, the public will have no reason for believing anything they say. It does not matter how many earnest and honest-seeming people announce that, as expected, nothing was found behind the door of the Gantenbrink Shaft, public opinion will not be fooled, for the Egyptologists have missed their chance to be trusted.

The old Roman historian Cornelius Tacitus (AD 55–120) said it all: 'Those who dislike criticism show that they deserve it.'

NOTES

1 Däniken, E von, *The Eyes of the Sphinx*, Berkeley, 1996
2 Sasse, T, 'Der Schacht des Cheops', in *GRAL*, No 5, 1993
3 Goyon, G, *Die Cheops-Pyramide*, Bergisch Gladbach, 1979
4 Haase, M, 'Wp-w3wt. The one who opens the ways', in *GRAL*, No 5, 1993
5 Schüssler, K, *Die ägyptischen Pyramiden*, Cologne, 1983
6 In the *Daily Telegraph*, 7 April 1993

7 Reuter telex of 16 April 1993
8 'The great pyramid mystery', in the *Mail on Saturday*, 17 April 1993
9 'Secret chamber may solve pyramid mystery', in *The Times*, 17 April 1993
10 Al-Makrizi, Taki, *Das Pyramidenkapitel in al-Makrizis 'Hitat'*, translated into German by E Graefe, Leipzig, 1911
11 Tompkins, P, *Cheops*, Bern, 1975
12 Däniken, E von, op cit
13 'Chephren-Pyramide – Fluch des Pharaos', in *Der Spiegel*, No 33, 1969
14 Yoshimura, S et al, *Non-Destructive Pyramid Investigation by Electromagnetic Wave Method*, Waseda University, Tokyo, 1987
15 'Sphinx, Riddle Put to Rest?', in *Science*, Vol 255, No 5046, 14 February 1992
16 West, J A, *Serpent in the Sky*, Eheatin, 1993
17 Wölfli, W et al, *Radiocarbon Chronology and the Historical Calendar in Egypt*, reprinted from *Chronologies du Proche Orient*, BAR International, Series 379, Paris, 1987
18 Fischer, J, 'Noch immer Spekulationen um eine Geheimkammer in der Cheops-Pyramide', Report 515 DPA 0185 of 27 June 1994 from Cairo
19 Sasse, T, Interview with Professor R Stadelmann on 15 June 1993 in Berlin
20 Wildung, D, 'Pharaohmarkt, Technik der Pyramidenmystik', in *Frankfurter Allgemeine Zeitung*, 5 August 1993
21 Keys, D, 'Discovery at pyramid was accidental', in *The Independent*, 16 April 1993

Index

Abel 32
Aborigines 93
Abraham 47f, 50f, 56f
Adam 22f, 25f, 30, 40, 75, 80, 104
Adonis 11
Africa 113
Africanus, Julius 34
Ahmadiyya movement 66
Ahura Mazda 72, 74
Aldebaran 25f
Allah 71, 99
Alvarez, Luis 152
Angels 21, 22f, 30, 35, 38, 43f, 45, 47, 50, 54f, 72, 77, 88, 94, 96f, 105f
Apocalypse 62f, 72, 106
Archaeology 137, 139
Archaeopteryx 118
Archangel 18, 23, 31, 97, 106
Arishtanemi 79
Artaxerxes 72
Assurbanipal 29, 89
Astronomy 137
Atom 81f
Atra Haris 36
Australia 93
Austria 124
Avesta 72, 73
Babylon 36
Baruch 33
Bayraktutan, Salih 37
Beckwith, Steven 139
Bel 89
Benares 86
Berlin 149
Bern 103
Bhagavata-Purana 75
Bible 20, 28f, 32f, 35, 42, 52, 78
Böhl, Franz 47f
Book of the Patriarchs 2, 5, 7f
Borsippa 90
Boston 126, 137, 153, 154
Brahma 83
Brain 119, 121f, 124, 129, 134f,
Brazil 131
Buddha 85f, 103
Buddhism 67, 77, 85
Cain 32
Cairo 150f, 153
California 136
Cayce, Edgar 64, 66
Charon, Jean E 66
Cheops 148, 151f, 160f, 162
Cheops Pyramid 144, 146f, 153, 155f, 161, 162f
Chephren 153f
Chephren Pyramid 152
China 92, 113
Chronos 80
Codex Hammurabi 88f
Codex Sinaiticus 19
Codex Vaticanus 19
Consciousness, cosmic 135
Cortes 93
Cosmos 139
Daniel 63, 68f, 105
Darius the Great 72
David 33, 67
Day of judgement 57, 61f, 68, 70, 99
Dead Sea Scrolls 90
Death 61, 91, 94
Delitzsch, Friedrich 20
Deva-Yuga 74, 87
Dinosaurs 117f
Diodor 80
Dixon, W 147, 159f
Dobecki, Thomas 155
Dormion, Jean-Patrice 153
Dreyer, Günter 150
Drona Parva 76
Ea 86
Eddington, Arthur 82
Egorov, Christel 150
Egypt 21, 34, 49, 80, 117, 144f
Einstein, Albert 110
Eisenmenger, Dr 28
Electron 82f, 103
Enlil 86
Enoch 24, 51f, 63, 77, 80, 84, 88, 90f, 92, 94f, 104f, 144, 151f
Enuma Elish 36
Enzymes 116
Ethics 18, 47, 102
Europe 136, 153
Eusebius 34
Eve 26f, 30f, 40, 75, 104
Evolution 99, 103, 117, 118, 135f
Extraterrestrials 39f, 43, 57, 98f, 104, 109f, 120, 122, 125, 129f, 133f, 135f, 138f, 144f
Ezekiel 48f, 69
Fallen angels 32f, 43, 57, 90, 94
Farmer, Jack 136
Fasold, David 37

Fiebag, Johannes 124f
Fischer, Jörg 156
Flindt, Max 119
Flood 29, 32f, 36f, 43, 52f, 55f, 79f, 90, 135, 152
Fogg, Martyn 136
Fossil research 113
Fraknoi, Andrew 136
France 131
Frankfurt 120
Gabriel, Ulrike 120
Gaganacara 76
Galilei, Galileo 73
Gandhi, Mahatma 50
Gantenbrink, Rudolf 145f, 148f, 149f, 156f, 160, 162f
Geographos 11
Germany 124, 131
Gesar 91f
Gilgamesh 29, 36
Giza 148, 153. 155. 157
Glasenapp, Helmut von 81
Gluskabe 92
God 18f, 21f, 26, 32, 37f, 52, 56f, 64, 70, 76f, 89, 93, 96f, 99, 102, 106
Gohed, Amr 152
Goidin, Gilles 153
Golden age 74f
Gospels 19
Greece 87
Gressmann, Hugo 93
Gujarati 72
Gyelrap 92
Haase, Michael 149
Hammurabi 88f
Hawass, Zahi 155
Heidelberg 139
Hephaistos 80
Heraclitus 87
Hermes 11, 151
Herodotus 162
Hesiod 41, 80
Hieroglyphs 160f
Himalayas 86
Hinduism 74f
Hitat 52, 151f
Homer 41
Homo erectus 113
Homo sapiens 40, 43, 61, 99, 113
Hopkins, Budd 123
Horus 80
Hughes, David 139
Hybrids 129, 133f, 135
Ibn Battuta 152
Ideology 110, 114
Idris 144
Imam 71
India 65, 72, 77, 86, 103
Indians, American 92
Iran 72
Isaiah 67f
Ishtar 89
Isis 80
Islam 70f
Israel 69
Jacobs, David 125
Jainism 77f, 82f, 86, 103
Japan 113
Jaredites 36
Java 113
Jaynes, Julian 118
Jehrameel 47
Jeremias, Alfred 86
Jerusalem 20, 100
Jesus 19f, 31, 67f, 70, 88
John 63, 70, 105f
Jupiter 73
Kala 80
Kali-Yuga 87
Kalpas 85
Kandshur 92
Karma 77, 81f, 83, 93
Kehl, Robert 20
Kenyon, Kathleen 49
Ketumati 86
Keys, David 162
Khecara 76
Klein, Michael 136
Koran 37f, 62, 67, 70f
Krishna 77
Krtayuga 74
Kujundshik 29
Kukulkan 92
Küppers, Werner 68
Lamech 56, 90f
Landmann, Leo 69
Lehner, Mark 154
Licharev, Konstantin 103
Linz 120
Logic 128f
London 136
Lorber, Jakob 65, 66
Ludwiger, Illobrand von 138
Luke 20, 62
Mack, John E 126f, 128, 132, 138
Mahabharata 76
Mahadeva 76
Mahaparinibbana-Sutta 86
Mahavira 79, 84
Maha-Yuga 87
Mahdi 71f, 93, 99
Maitreya 86
Manetho 34, 80
Marduk 89
Marino, Lori 136
Mark 20, 61f
Mars 16, 73, 111, 119, 136
Mary 88

Matthew 20
Mayas 81, 92
Mercury 73
Messiah 67f, 84, 93f, 100f
Methuselah 52f, 56, 80, 90, 94f, 144
Mexico 93
Midrashim 50
Mohammed 71
Moon 16f, 73, 98, 111, 119
Morality 18, 32, 40, 47
Mormon 35f, 67
Moses 89
Mythology 43, 86, 91, 102
Nabu 89
Naukratis 21
Nephi 36
Nestroy, Johann 137
New York 136
New Testament 20, 50, 61, 63, 65
Nimrod 50
Noah 24f, 29, 36f, 90, 135
Noah's Ark 35f, 37f
Old Testament 52, 54f, 65, 90, 102, 105
Original sin 37, 43, 67
Original texts 19f, 68
Osiris 80
Padmasambhava 92
Palaeo-seti philosophy 39, 44f, 48, 54, 94, 99, 104, 138
Papagiannis, Michael 137
Parsees 72f
Peru 93
Pizarro 93
Planck, Max 83
Plato 21
Plutarch 34
Pope 133
Quadriga solis 73
Radiation 116
Raju, Satyanarayana 65
Ray, Tom 120
Reincarnation 77, 78, 83
Reptiles 118
Rigveda 76
Rishabha 79
Sabhaparva 76
Sai Baba 65f, 103
Samas 86f
Sapphire stone 22f
Sasse, Torsten 146, 160
Saurid 144
Schoch, Robert 153, 154f
Schomerns, H W 69
Script 22, 121
Sebennytos 34
Sekhemkhet 163
Seth 23f
Sicily 80
Sin 86f
Smend, Rudolf 48
Smith, Joseph 35
Socrates 21f
Sodom and Gomorrah 32
Soul 83
South America 36
Sphinx 153f
Stadelmann, Rainer 146, 150, 157f, 159f, 160f
Stadler, Beda 103
Strabo 28
Struwe, Maria 124f
Sumerians 36, 87
Sun 73
Susa 89
Switzerland 124
Tacitus, Cornelius 164
Tandshur 92
Theory of evolution 99
Theuth 21
Thunder-hammer 91
Thutmosis IV 154f
Tibet 91f
Time machine 134
Time travel 134
Tiphon 80
Tirthamkaras 79, 84f, 86, 103
Titus 20
Tokyo 153
Tutankhamun 163
UFO 26f, 122f, 131f, 138f
Universe 88, 136
Upuaut 146f, 161
Uruk 29
Utnapishtim 29, 135
Vaihayasu 76
Vanaparvan 76
Veda 76
Venter, Craig 114
Venus 73
Vimanas 76, 85, 99
Vishnu 77
Wabanaki 92
Washington 122
Wellnhofer, Peter 118
Wertz, James 136
West, John Anthony 154
White, John 130
Wildung, Dietrich 162
Wölfli, W 155
Xerxes 72
Yugas 86f
Zarathustra 72, 74
Zurich 155